ADVANCE PRAISE FOR

Decolonizing Environmental Education for Different Contexts and Nations

"About a decade ago, educators began to realize that Indigenous peoples traditionally live the gold standard of sustainability. It soon eclipsed the Eurocentric, three-pillar (economy, society, environment), hegemonic, sustainability agenda favored by UNESCO. As Marie Battiste rationalized, 'You can't be the global doctor if you're the colonial disease.'

A decade later, we are gifted by a book that presents a diverse sustainability agenda aimed at ecological and social justice in environmental education (EE). Its diversity is secured by conveying experiences from eight countries (Australia, Bangladesh, Canada, Brazil, Ghana, Jamaica, Mexico, and USA). Each animates important conceptual and methodological contributions to EE scholarship.

Just as diversity breeds ecological strength in natural settings (aka Mother Earth), this book's diversity offers a refreshingly strong array of knowledge and wisdom. It maps out an evolved agenda for sustainability in EE, which promises to eliminate the current Euro-American global transculturation—the ideology of colonialism. Megan Bang calls such an innovation a 'move toward just, sustainable, and culturally thriving futures.'"

—Glen Aikenhead, Emeritus Professor, Curriculum Studies,
University of Saskatchewan, Canada

"We live in challenging times! We look to leaders to share seeds/stories that enable our reworlding so that we co-design better ways, generate new-normals, and actively provoke diversity in our ways of knowing and being. The authors of this book have listened and offer us insight into possibilities and opportunities for co-creation of socio-ecological justice, should we be brave enough to act. Who better to encourage change towards decolonized, hopeful, inclusive futures than environmental educators: leading us beyond Western entitlement (re)storying our education practices with our ecosystems in mind."

—Peta White, Senior Lecturer, Education (Science Education),
Deakin University, Australia

"This book offers an exciting contribution and presents creative experiences and reflections that led the readers to get in contact with different perspectives addressed to decolonize Environmental Education field. This is a very successful attempt to present one opposite view of what has been considered as settler-colonial projects from the Western worldview, which reinforces the 'human exceptionalism and supremacism'. Ultimately, the purpose of this book is to (re)story the field as 'a site of hope' through educational practices meant to construct particular experiences based on/for socio-ecological justice for humans and others-than-humans. Based on different points of origin, from different land educational contexts, and reflecting from different conceptual backgrounds, the intention is not to present 'static definitions or final and absolute answers to questions' to environmental education practices, but to offer some possibilities to guarantee that humans and 'others-than-human entities also have the right to exist, to live well, in healthy ecosystems'."

—Luiz Marcelo de Carvalho, Professor, Education Department, Institute of Bioscience,
University of State of São Paulo, Rio Claro Campus, Brazil

"The need for this book is essential. This book provides needed respect for world cultures, ethnicities, regions, religions, politics, genders, sexualities, and views regarding Environmental Education. The authors break down barriers of ethnocentrism, specifically the White and Western view of Environmental Education, and open an inclusive dialogue to advance the conversation forward in an interconnected way. Provided in the chapters are thought-provoking insights that elicit movement in a positive, collaborative direction by changing the current narrative. The diversity of experience, ideas, and contexts presented help to build a global community of change-makers to stop a dire environmental future. The wisdom in this book creates an opportunity to learn, grow, and make a significant change on a global scale and passionate commonality for the environment through education."

—Shannon McLean-Minch, Elementary School Principal, Community Montessori, Boulder Valley School District, Boulder, Colorado, USA

Decolonizing Environmental Education for Different Contexts and Nations

(Post-)Critical Global Childhood & Youth Studies

Márcia Aparecida Amador Mascia,
Hongyan Chen, Silvia Grinberg and Michalis Kontopodis
Series Editors

Vol. 3

The (Post-)Critical Global Childhood & Youth Studies series
is part of the Peter Lang Education list.
Every volume is peer reviewed and meets
the highest quality standards for content and production.

PETER LANG
New York • Bern • Berlin
Brussels • Vienna • Oxford • Warsaw

Decolonizing Environmental Education for Different Contexts and Nations

Kathryn Riley, Janet McVittie,
and Marcelo Gules Borges, Editors

PETER LANG
New York • Bern • Berlin
Brussels • Vienna • Oxford • Warsaw

Library of Congress Cataloging-in-Publication Control Number: 2021054900

Bibliographic information published by **Die Deutsche Nationalbibliothek**.
Die Deutsche Nationalbibliothek lists this publication in the "Deutsche
Nationalbibliografie"; detailed bibliographic data are available
on the Internet at http://dnb.d-nb.de/.

ISSN 2297-8534
ISBN 978-1-4331-9174-9 (paperback)
ISBN 978-1-4331-9183-1 (ebook pdf)
ISBN 978-1-4331-9184-8 (epub)
DOI 10.3726/b18830

© 2022 Peter Lang Publishing, Inc., New York
80 Broad Street, 5th floor, New York, NY 10004
www.peterlang.com

To all the entities that call this planet home

Table of Contents

List of Figures

Note from the Series Editors

We are very pleased to introduce *Decolonizing Environmental Education for Different Contexts and Nations*, the third volume of the book series (Post-) Critical Global Childhood & Youth Studies. This edited volume pushes us to expand existing theories and approaches to environmental education and to consider new possibilities for exploring, understanding and embracing situated knowledges across diverse contexts in countries such as Canada, Australia, Brazil, Ghana and Bangladesh. This investigation supports an important debate on our *mutually entangled futures*—in Kathryn Riley's words, which is crucial at a time when environmental and public health crises are destroying the livelihoods of those least responsible for these. As the authors of this volume argue, there is an urgent need to trouble and challenge the Western worldview that dominates science, society and education in the twenty-first century.

We hope that this third volume of (Post-)Critical Global Childhood & Youth Studies highlights the importance of the exchange of ideas on contemporary issues affecting children and young people around the world while exploring possibilities for local and global social change, which is indeed the focus of the Series. The Series encourages novel approaches to co-producing knowledge in fields such as: urban, rural and Indigenous childhood & youth; children's rights; social policy, ecology and youth activism; faith communities; immigration and intersectionality; mobile Internet, digital futures, and global education. It discusses the geopolitics of knowledge, and decolonial and anthropological perspectives, among others. (Post-)Critical Global Childhood & Youth Studies is addressed to relevant scholars from all over the world as well as to global policy makers and employees at international organizations and NGOs. This book is indeed an invitation to draw upon such different fields of expertise and areas of activity, and to co-develop a different environmental education from within the most diverse cultural and geographic contexts.

São Paulo, Shanghai, Toronto, Buenos Aires & Leeds, October 2021
Series Editors

Acknowledgments

This book is a collective process of decolonizing ourselves. It emerged as a result of sharing our experiences, listening, and standing in solidarity with Others, despite our often-diverging perspectives as situated in different contexts and nations. *Decolonizing Environmental Education for Different Contexts and Nations* describes individual and collective paths of reflection, animated by history and stories that connect environment and education through, and from, different lands and territories. Over the last two years we have organized meetings and have exchanged thoughts, materials, arguments, challenges, and hopes through our work and actions in our contexts. Based on these conversations, we connected contexts and nations, proposing not to produce a guide on how to decolonize environmental education, but, above all, make explicit our effort to change priorities and decenter human cultures in, and for, environmental education. Most of the authors in this book contributed throughout the meetings, others contributed through their connections with authors in the book.

First, we would like to acknowledge the work from Márcia Mascia, Silvia Grinberg and Michalis Kontopodis in editing this series. We are especially grateful to Michalis Kontopodis for helping us with our book proposal and moving us forward to conclude the book. Our thanks to Juliano Camillo for first connecting ideas of this book with Peter Lang. Thanks also to Dani Green, our primary contact at Peter Lang.

We are grateful to all of you from our contexts that helped us in the process of decolonizing our philosophies, theories, and practices: Rodrigo Ramirez (in memoriam), Buddy, the grey wolf, the doctoral research award from the International Development Research Centre (IDRC), the Dr. Rui Feng Geological Sciences Graduate Studies Award (University of Saskatchewan, Canada), and The University of Saskatchewan Ph.D. Fellowship. Thank you to the Laitu Khyeng Indigenous Elders, knowledge holders, leaders, and youth participants and to the Campesinas and Campesinos for your warm welcome into your communities and for patiently sharing your stories and knowledge about sustainability and land and water management. Thank you to Marla Guerrero Cabañas, the coordinator of the Calakmul field team of Fondo para la Paz for your trust, openness, time and friendship, and the facilitators and community technicians of Fondo para la Paz: Reynaldo Zepahua Apale, Ezequías Hernandez Torres, and Gelasio Maldonado de Paz for your accompaniment, kindness, patience, and humor. Thank you to the community of La Loche for your trust in the education of your youth, and for your compassion, humor, forgiveness, and kindness. Marci Cho!

We are also grateful for the authors and artists that contributed to this book. We would like to especially thank Juliano Camillo, Zuzana Morog, Vince Anderson, Sky McKenzie, Araceli Leon Torrijos, Alice Johnston, Kylie Clarke, Roseann Kerr, John B. Acharibasam, Elsa McKenzie, Ranjan Datta, Julio Karpen, and Kai Orca. We also thank Jebunnessa Chapola for her insightful contributions to our conversations. We also extend our deepest thanks and love to our families: Trevor, Zuzana, and Charlie Morog, Erin and Natalie McVittie, Iain, Ragnar, Rowan Phillips, and, also, Daisy, Francisca and Samuel Sampaio, and the four-legged friends that make our family complete. Lastly, but certainly not least, we extend our heartfelt gratitude to the Earth that sustains us: the critters of land and sea, the energies, sky, land, waters, and First Peoples that teach us there is a more gracious and reciprocal way of living on, and with, this finite planet.

Abbreviations

CaC: Campesino-a-Campesino
CHT: Chittagaong Hill Tracts
DW-K: Dominant Western Knowledges
DWW: Dominant Western Worldview
ECEE: Early Childhood Environmental Education
EE: Environmental Education
EfS: Education for Sustainability
ESD: Education for Sustainable Development
IK: Indigenous Knowledges
KG2:   Kindergarten 2
LEdC: Licenciatura em Educação do Campo
MST: Landless Workers Movement
NAAEE: North American Association of Environmental Education
PHG: Prairie Habitat Garden
RCAP: Royal Commission on Aboriginal People
TC: Community Time
TEK: Traditional Ecological Knowledges
TRC: Truth and Reconciliation Commission
TU: University Time
UFSC: Federal University of Santa Catarina
UCC: United Church of Christ's Commission for Racial Justice
UN: United Nations
UNEP: United Nations Environment Program
UNESCO: United Nations Education, Scientific, and Cultural
            Organization

Decolonizing Environmental Education for Different Contexts and Nations

JANET McVITTIE

Be good to the planet. It is where I keep all my stuff. (Anonymous)

Introduction

Environmental Education (EE) as a discipline has its roots within the Dominant Western Worldview (DWW). The release of Rachel Carson's book, *Silent Spring*, in 1962 (1962/2002), affected the way White, Western people thought about humanity's interventions in ecosystems. Until then, White, Western people acted, in general, as if they were the only members of the only species that mattered, and the planet's resources were for their consumption.[1] This relatively small and wealthy set of humans did not include even those marginalized within their nations, such as Indigenous peoples; it certainly did not include those from outside of Euro-America (which includes those parts of North and South America where people of European ancestry form the majority of the population), Australia, and New Zealand. Interestingly, now that more researchers are examining Indigenous Knowledges (IK), there is more clarity about what it means to live sustainably, in recognition that the planet is finite. Carson's book should have pushed people to examine their world views, but this examination has taken some time.

Seawright (2014) argued that DWW (he used the term "settler epistemology") has created injustices. He noted: "The dominant epistemology of settler society provides racialized, anthropocentric, and capitalist understandings of place" (p. 554). He goes on to argue that without an understanding of settler epistemology, a person will be unable to resist appropriately. DWW is characterized by the belief that humans (particular humans—male, heterosexual, White) are superior to all other entities (an anthropocentric view), that the planet and its resources are here for the particular use of these humans, that land should be privately owned (a particular focus of Seawright's is on the claiming of land), that capitalism and democracy are the only appropriate approaches to human politics and economy, that the economy must be constantly growing, that most things can be organized into dichotomies—good & bad, female & male, black & white, night & day, nature & culture, wild & domestic, etc. Responding to these binary logics, Kai Orca, in Chapter 9, focuses on the damage done by dichotomies, connecting the violence done to children by assigning gender to them at birth, to the violence done to nature by naming it as separate from humans. Further, Kathryn Riley, in Chapter 2, draws on posthumanist perspectives to generate an interactive relationship, an assemblage, with Buddy the grey wolf. In this way, she works to (re)configure oppositional and dualistic difference and (re)draw subject/object boundaries, while also challenging the subsequent hierarchical ordering based on this difference.

The authors of the chapters in this book are a group of educators with a common interest in environmental education (EE), coming together through different wayfaring routes from different points of origin, including Africa (Ghana), Asia (Bangladesh), Australia, North America (Canada, Mexico, United States), and South America (Brazil). As a multi-dimensional group, we are learning about different places and contexts. In all locations we derive from, there are social and ecological issues that need to be resolved.

There is current research from every location on implementation and uptake of EE, and on how one might make the locale more socially just, but generally, a DWW definition of and approaches to EE have informed what has been done. As a group, discussing what the issues were/are in each context/location, and how different approaches had been tried and have worked to varying degrees (some have worked not at all!), we realized that the location—geography, history, politics, culture—affected the ways in which EE was implemented, and the results that were derived. The concept of globalization postulates that all nations are interconnected through economic, cultural, and political relationships. Indeed, all nations exist on one planet, as an autopoietic system with finite resources. Although the actions in one context affect other contexts, the solutions that work in one context do not necessarily work everywhere. In this book, we authors focus mostly on cultural connections and differences. We note that individual countries (although unique in many ways) in a globalized world tend to slip into the gravity well formed by DWW. A DWW form of EE is not appropriate in every locale.

To support separating or delinking (Mignolo, 2011) from DWW, decolonization became a primary base for our thinking to inform how best to implement EE. Most of the authors use the term decolonization, meant in this book as a method to address a long-standing history of Europe's incursion into the Americas, Africa, Australia and New Zealand, and parts of Asia. Each chapter in this book explores a unique context, and each author takes up their own conceptual framework, but in each case, the authors move forward understandings of EE through decolonization. In this chapter, the introduction, we present the research literature discussion for different terms and concepts to further the readers' understanding for the rest of the book. We begin with a brief synopsis of "Environmental Education" (EE), go on to examine "Decolonization", then how different knowledge systems can (or not) be integrated in a discussion on "Hybridity", and "Border thinking". We complete this chapter with a focus on Indigenous peoples, in which we explore meanings of IK.

Why Read This Book?

The year 2018 saw the publication of two change making sets of information. The Intergovernmental Panel on Climate Change (2018) released a report, noting that if planetary temperatures rise higher than 1.5C above pre-industrial levels, there will be significant and irreversible loss of human habitats and foods. For humanity, the report means we must make significant changes in how we operate, such that "Global net human-caused emissions of carbon dioxide (CO_2) would need to fall by about 45 percent from 2010 levels by 2030, reaching 'net zero' around 2050". There can be no hesitation in our actions. We must reduce our use of materials that create greenhouse gases. The other significant report addresses extinction rates for living species, with the Centre for Biological Diversity (2018) suggesting the current rate is 1,000 times the background rate. The World Wildlife Federation (2018), on the other hand, noted the rate is somewhere between 1,000 and 10,000 times the normal background rate. The differences in extinction rates are due to lack of knowledge of how many species currently exist. Within the range of estimates, 30–50% of species will be gone by 2050, with pollinators being significantly affected.

Since the reports, predictable outcomes have occurred, direct fall-out of human actions affecting the environment. Planetary systems have been disrupted by the global pandemic caused by a virus making the leap from its animal host to humans, by catastrophic forest fires in Australia, the Amazon, and North America, as well as by extreme weather events.

Surely all of this will make a difference in how humans interact with the natural world! But we have been hearing about climate change since the

1970s, the extinction crisis since the 1960s, and those humans with the most power to enact change, generally, have not acted. Somehow, the severity of the crisis must be communicated to the whole of humanity. There can be no more "business as usual". Fundamental changes in how humans relate to the environment are required.

Rectifying the damage done, though, is a wicked problem. Finding a solution to one part of the problem can often create more problems in another situation, in another place. There needs to be either a wholesale change in how humanity interacts with the rest of the world, or everyone can take up the changes they can make in their locales. In either case, understanding DWW and selecting what is useful from it and modifying or replacing other parts of it are necessary.

For the authors of this book, the issue is greater than humanity's survival in the face of climate change. We care deeply about the land and other-than-human entities. We all believe that other-than-human entities have inalienable rights to exist, to live well, in healthy ecosystems. This challenge to the anthropocentric view of the land is a challenge of one aspect of DWW. Humans, specifically white heterosexual male humans, are not the only entities which count. In this book, most of the authors look to IK as potentially leading us forward to live in better ways with the planet, home to millions of species of living entities. The reason we look to IK is outlined by Dei (2000), as he cited a paper that he co-authored in the same year:

> More specifically, the term/notion 'Indigenous' refers to knowledge resulting from long-term residence in a place (Fals Borda, 1980). Roberts (1998: 59) offers a clear conceptualization of 'Indigenous' as knowledge 'accumulated by a group of people, not necessarily Indigenous, who by centuries of unbroken residence develop an in-depth understanding of their particular place in their particular world'. 'Indigenous' signals the power relations and dynamics embedded in the production, interrogation and validation of such knowledges. It also recognizes the multiple and collective origins as well as collaborative dimensions of knowledge and affirms that the interpretation or analysis of social reality is subject to differing and sometimes oppositional perspectives (Dei et al. 2000). (p. 114)

In taking up various forms of decolonization, most authors in this book explore the importance of bringing commensurate attention to IK (the knowledge, the values, the ontologies) in EE. This acknowledges that Indigenous peoples have developed sustainable ways of knowing, being, thinking, and doing over extended periods of time, living with and on the land, in ways that ensured healthy ecosystems and a vibrant sense of relationality among all entities.

Thus, the purpose of this book is to decolonize (Andreotti, 2015; Hunt & Holmes, 2015; Battiste, 2017; Dei, 2000; Mignolo, 2011) EE, especially as it pertains to land (Calderon, 2014; McCoy, 2014; Paperson, 2014; Seawright, 2014; Simpson, 2011, 2014; Tuck et al., 2014), through examination of different ways of knowing (with) the world (Cajete, 2005; Dei, 2000; Deloria, 2001; Kayira, 2013; Musopole, 1994; Seawright, 2014; Sindima, 1990, 1995), so as to support ecological and social justice in different contexts and nations. As noted above, each of the authors in this book derives from a unique locale, context, culture. Each country offers a unique system regarding land ownership and distribution, and resources that are exploited. Even within countries, there are unique contexts. For example, of three authors writing from Canada, Vince Anderson added to EE theory through integrating social and ecological justice, Alice Johnson taught children in a remote location, and Janet McVittie and Marcelo Gules Borges taught adults in a teacher education program. Thus, even within one country, contexts varied widely. Further, several authors in this book do not specifically address IK, one because she viewed knowledge as hybridized, and one because she is not Indigenous and is therefore unwilling to take up IK. Nonetheless, all authors examine decolonization of EE through their own conceptual and contextualized understanding of what this means for themselves and their philosophical, theoretical, and practical contexts.

We believe the process we went through, and the different approaches to decolonization that are taken up, will support others in decolonizing EE, as appropriate, for their contexts. We hope that by delivering EE in culturally and ecologically appropriate ways for each context, this book can support humans in becoming passionate stewards of the environment.

A Brief Synopsis of Environmental Education

Environmental education has its roots far deeper than when the term was first seriously taken up by Stapp (1969) and United Nations Environmental Program (UNEP), but the focus has varied over time, in different locations, and for different political ends. Nature education, nature play, camping education, place-based education, Education for Sustainable Development (ESD), and Education for Sustainability (EfS) are just a few of the terms that are or have been used within EE.

During the late 1700s, Pestalozzi promoted schools for young children whose parents could not otherwise afford education, or indeed, were often not even able to watch over them. Pestalozzi was inspired by Rousseau (1762/2016), who considered human society to be corrupt, and so wanted to get children into nature. Over the next few centuries, there were pockets of interest in getting children outdoors to learn. Sometimes the focus for getting children outdoors was on the child and curricular content (such as from the Nature play movement, which, as Biedenweg et al. (2013) noted was driven by a concern that "children were missing vital interactions with the natural world that could shape their future careers" (p. 10)). Sometimes the focus was on changing the pedagogy, such as for Dewey (1938) and Schwab (1963) who required experience and inquiry as pedagogical methods. Both of these scholars believed in the value of the milieu, getting children into their worlds outside of the classroom, so the children could interact with the issues their learning would focus on solving. Teaching practices, thereby, might take children to nearby urban areas, or more natural areas. Learning or morality might be the focus. Or Carson (1965/2017) argued for children to go into natural areas, so as to experience a sense of wonder. Recently, there has been much research on the health and development benefits for children (see Maller et al., 2006, for an early review of this work).

There was a shift in EE in the late 1960's when the UN, as a body, noticed that environmental problems were indeed a global issue. Bill Stapp (1969) was integrally involved. Stapp (and others) (1969) argued: "man [sic] is an inseparable part of a system, consisting of man, culture, and the biophysical environment, and that man has the ability to alter [those] interrelationships" (p. 31). The United Nations conference in Stockholm (1975) created the UN Environmental Program (UNEP) within the Education, Scientific, and Cultural Organization (UNESCO). UNESCO/UNEP hosted a conference in Belgrade which resulted in the goal for environmental education being stated:

> To develop a world population that is aware of, and concerned about the environment and its associated problems, and which has the knowledge, skills, attitudes, motivations and commitment to work individually and collectively toward solutions to current problems, and the prevention of new ones. (UNESCO/ UNEP, 1975, p. 3)

The 1977 UNESCO/UNEP Final report (from the Tblisi meeting) noted the connection amongst social issues, economic, political, and ecological issues, and reiterated the need for humans to attend to the environment: "to create new patterns of behaviour of individuals, groups, and society as a whole towards the environment" (UNESCO/UNEP, 1977, p. 26). The shift from a focus on individuals growing up to be productive members of a scientifically based society in science education, to a focus on the environment, while recognizing the complex interplay amongst human politics, cultures, economies was an integral shift for education. Although there was no specific statement that the approach should be ecocentric, there was an opening for this to happen. As Seawright (2014) noted, it is difficult to change one's beliefs and actions without understanding the epistemologies that drive those actions. Similarly, environmental educators, without knowing their current epistemologies, would tend to continue with an anthropocentric approach. Consistent with DWW, when the term EE was first coined, it tended to be taught within science. EE was often one topic, perhaps taking up one quarter of a secondary biology course. Thus, the EE movement (using the term EE) was initially taken up through the study of ecological principles, and almost always, these were learned from a book inside the walls of a classroom and presented within DWW.

This is not to say that science education is not important. Rather, as we present different approaches to EE in this book, we suggest that there can also be other curricular homes for EE. Certainly, ecological principles are important, and humans have learned much through the systematic study of ecosystems, but, as Orr (2010) noted "all education is environmental education" (p. 242). If we do not teach about the environment, we are saying the status quo is just fine; when we do teach about/in/for the environment, we are demonstrating the environment has value—good or bad value would depend on the approach. Capra and Luisi (2014) promote "systems thinking", which recognizes that all entities are connected, all are part of one large system. When one entity fails, other entities are affected. Orr (2010) and Capra and Luisi (2014) present, contrary to the anthropocentric approach of other environmental educators, an ecocentric approach. The land and all other entities have inherent rights and value. For these scholars, the focus should be on ecosystems, not on humans. Further, formal schooling is not the only place in which EE can be learned. For example, Roseann Kerr (Chapter 5), addresses non-formal education for Campesinos/as in Mexico; people create wisdom as they work collectively towards solutions.

Since the emergence of EE, there has been critique. For example, Martusewicz et al. (2011) argued that environmental educators often adopt a materialistic scientific approach of data collection, then of framing regulatory legislation or production of institutional procedures, as the means to resolving environmental problems. Tilbury (1995) argued that many communities were left out of EE; a response was to revise the overall goal of EE to include Education for Sustainable Development ESD, so as to include less affluent countries. Davis (2009) argued that EE must look to the diversity of Indigenous cultures for new directions.

About two decades after the Tbilisi report (UNESCO/UNEP, 1975), the Brundtland Commission (UNESCO, 1987), through engaging with minoritized communities such as Indigenous peoples, women, and youth, as well as drawing on local authorities, outlined a plan for "sustainable development". This returned the focus to humans, away from the environment. See Kopnina (2012) for a thorough critique. Contained within the UNESCO (1987) document, *Our Common Future*, is a statement supporting "policies that will sustain and grow the environmental resource base. And we believe such growth to be absolutely essential to relieve the great poverty that is deepening in much of the developing world" (p. 11, item 3). Although this sounds as if the UN is speaking to the global poor, the authors clarify that sustainable development does imply limits, but not absolute limits. Rather, there are limitations imposed by the present state of technology and social organization on environmental resources and by the ability of the biosphere to absorb the effects of human activities. The report argues that technology and social organization can be both managed and improved to make way for a new era of economic growth (UNESCO, 1987, p. 16). From the report, the definition of ESD is clarified: "Sustainable development seeks to meet the needs and aspirations of the present without compromising the ability to meet those of the future" (UNESCO, 1987, p. 39). And the focus is definitely anthropocentric.

ESD required reconciliation between environmental conservation and economic development (Tilbury, 1995). From the onset, there has been criticism of ESD from environmental educators. Blum et al. (2013) noted that the tension between EE and ESD often comes down to the goal of education itself: is education about learning content, or for developing capacity to think more critically? They raised the question regarding the preposition *"for"* in ESD being an issue for many EE researchers, in that educating "for" something, implies that students will be indoctrinated to live in particular ways. However, a bigger concern for ESD was the belief that development could continue, without creating environmental damage. As new technologies are developed, the environment does not necessarily improve, despite best intentions. Two simple and obvious examples are: one of the factors in choosing automobiles over horse drawn carriages was that the air was so much cleaner than with horses and carriages, since the dust from horse droppings no longer went into the air; new low energy consuming LED television screens have led to North Americans purchasing more and bigger screens, resulting in increases in overall energy consumption.

In our broad review of environmental education, we encountered a variety of tensions:

- Is EE oriented towards human goals (anthropocentric), or is it about living respectfully with/for the environment (ecocentric)?
- Is EE to teach ecological principles, or to support children in developing a sense of wonder (as represented in Carson, 1965/2017), or to teach them to think critically (as suggested by Blum et al., 2013)?
- Is EE best taught as a subject area, with specific curricular outcomes, or in an integrative progressive approach (as described by Dewey (1938), and Jensen and Schnack (1997)) with curricular outcomes emerging as necessary?
- Is EE about teaching specific plans and approaches for bettering the planet (Education FOR sustainable development) or about supporting students in developing critical questioning and researching abilities (Wals & Jickling, 2000).
- Is science an appropriate curricular home for EE, whereby scientific principles can be learned, and can perhaps support technological improvements, or is EE a subject area of its own, which includes social-historical, cultural, communicative, as well as science concepts, or must it, as Orr (2010) argued, just be what we do and how we act in all living and learning situations? If EE is not addressed in a subject area, that sends a message that EE does not matter in those other content areas, and so should EE be incorporated across the curriculum (eco-musicology, for example)?
- Is nature enough? Will children, who are outdoors, becoming healthy and developing a sense of wonder, have the know-how and desire to solve the problems humanity has created?
- This begs the question of "what is nature?" Are humans natural, are we part of nature? If we believe not, what is the effect of believing in a nature/culture divide/dichotomy?
- Is there one solution for our complex wicked problem, or will each locale have to identify its own problems and work on them in ways that best suit their culture and environment?

These tensions highlight some divides in EE.

Although we authors believe that all humans should have access to personal health and safety, food security (and sovereignty), and economic security, we argue that the global rich have too much, and this has come at a cost to other species and their habitats. DWW promotes exploitation of the land; this is one of the main issues that must be challenged within DWW.

Numerous scholars and practitioners have realized that sustainability cannot be achieved without critical examination of social and cultural issues

(González-Gaudiano & Peters, 2008; Gruenewald, 2003; McKeon, 2012; Palmer, 1997). N. Gough (2013) noted that it is difficult to "think globally", but a promising approach would be to engage with scholars from other locales, to pay attention to their ways of knowing (see Ranjan Datta, Chapter 8, for an example of engaging with Indigenous people in Bangladesh). A. Gough (2013) desired to support environmental educators in creating partnerships with communities. Taking EE into the community means that schools and communities could examine social justice as well as ecological issues (see Marcelo Gules Borges, Chapter 3, for an example of a teacher candidate creatively integrating social justice into science education).

Sauvé (1996) noted that EE focused primarily on protecting the environment or nature for its aesthetic, economic, scientific, or other value without considering the associated human populations; UNESCO's thrust for ESD, she argued, did not explain a concept of environment, nor of sustainability, nor of education. However, ESD explicitly included social justice. Further examination of the relationships between social and ecological justice reveal they are integrally intertwined, as Vince Anderson, Chapter 1, demonstrates. Our task in this book is to show that when contexts—geographies, cultures, politics—are different, the approach to and process of EE can, and perhaps must, be adapted, inspired by the stories, experiences, and circumstances in each unique locale. By taking a deliberate decolonizing approach, and within our biweekly meetings, we authors shared writing as we critically examined the mainstream of EE and considered how to decolonize EE for our unique locales and contexts.

The point of this book is to contribute to, and expand, the conversation regarding how best to teach EE in varying local contexts; to illuminate how EE can, and perhaps should, change with culture, geography, history, politics, and economics.

Concepts the Authors Draw Upon

Each chapter in this book explores a unique context, and each author takes up their own conceptual framework, but in each case, they move forward the understanding of the role of decolonization in EE. As some of us took up IK, we were sensitive to the notion of the noble eco-Indian. As Grande noted:

> Some advocate the stereotype of the Indian-as-ecologically-noble-savage upholding "real Indians" as peoples with an inherent "at-oneness" with nature (e.g., Bierhorst 1994; Bowers 2003; Durning 1992; Hughes 1983; Piacentini 1993). Within this camp, an exoticized sense of indigenous peoples is perpetuated; Indians are held in fascination for their unspoiled, pure, and harmonious relationship with the land. (Grande, 2004, p. 94)

The belief in the "Indian-as-ecologically-noble-savage" moves the environmental problem from its actual source in DWW, to one where Indigenous peoples are co-opted to become settler-influenced environmental activists within settler approaches to land and ecosystems. Characterizing a whole people as "pure" reifies the root problem. Most of the authors in this book identify as of settler ancestry; we note our positionality, and struggle with tensions of decolonizing environmental education in our contexts acknowledging that IK is local and contextual. Dei (2000) noted that Indigenous peoples do not consider their knowledge to be universal. Indeed. Is there a universal knowledge?

Decolonization: Border Thinking and Hybridity?

We draw on decolonization for our theoretical framework. Decolonization is the process of undoing colonialism, however there are different ways in which places have been colonized (Tuck & Yang, 2012). This means, as Andreotti (2015) noted, that decolonization is a messy process, taken up in diverse and often contradictory ways. Although different forms of colonialism can be classified (for example: settler, exploitation, surrogate, internal), when Indigenous peoples in their places have been colonized, often multiple forms exist in one place. For example, in Canada, settlers have taken over the land (settler), exploited the resources (exploitation), and when Canada separated from Britain, the British background settlers took over control (internal). In a country such as Ghana, there were few European settlers. Nonetheless, with exploitation of resources (including the people as a resource for extracting material resources), and with the imposition of Anglo religions and school systems, the Ghanian people who had been put in charge became surrogate colonizers, such that even after British governors left, the colonized agenda continues; somehow, although the people have thrown off the foreign government, British systems continue. The case of Ghana illustrates the use of the term "postcolonial". It would seem that Ghana has been liberated from the colonizers, since by far the majority of people there are Indigenous, and Britain no longer controls the government. In this way, Ghana is now "postcolonial". However, the systems in place are still those of the British; for example, in the education system, children are indoors, in desks, in rows, learning (usually) in English, about the history of Europe. Thus, it is not just the existence of settler colonies that has meant a place has been colonized (Gabbidon, 2010). As well, the minds, bodies, souls, and knowledge of Indigenous peoples have been exploited, marginalized, modified. Dei (2000) noted: "as an African scholar in a Western academy, I see the project of 'decolonization' as breaking with the ways in which the (African) Indigenous human condition is defined and shaped by dominant Euro-American cultures, and asserting an understanding of the Indigenous social reality informed by local experiences and practices" (p. 113). He further argued, however, that different knowledge systems influence one another, that IK underwent evolution prior to contact, and that it is appropriate for knowledge systems to influence one another. Dei noted that IK is not lost, but rather is retained, within hybrid systems. These ideas are taken up by John Acharibasam (Chapter 6), as he explores the ontology and epistemology of the Kasena ethnic group in Northern Ghana, specifically focusing on the integration of Kasena's Traditional Ecological Knowledge (TEK) into Early Childhood Environmental Education (ECEE), drawing on guiding principles of "two-eyed seeing".

In this book, we do not only address colonization of minds. We are sensitive to Tuck and Yang's (2012) argument that "colonialization is not a metaphor". Although colonization of minds might be difficult to overcome, the theft of land must also be addressed. However, Grande (2004) noted a different argument about sovereignty over land. She stated that Indigenous (to the United States) peoples did not "own" land in their belief system. She argued that when Indigenous people take up the discussion of sovereignty, they are buying into Euro-American ways of thinking. What the Indigenous peoples believed was that the land was shared among all, all the entities, not just humans. Nonetheless, without a land base guaranteed to them in some way, Indigenous peoples are unlikely to gain any financial-, food-, home -security, or political power (see Janet McVittie and Marcelo Gules Borges, Chapter 7, for further elaboration). However, amongst Indigenous peoples, there are some interesting issues regarding land. For example, in both Canada and Ghana, there are numerous recognized tribal groups, whose beliefs about how to live with the land are very different from DWW concepts of private property, real estate, and exploitation (Seawright, 2014). However, these different groups might also be different from one another; they might occupy the same territory but draw on different resources. The question of sovereignty must be dealt with in different ways for different contexts.

To support decolonization, in our converstations we drew on Mignolo (2011) to consider epistemic disobedience as a process for delinking from colonial structures within EE. Mignolo noted that agents create and transform knowledge in response to their needs, as well as to institutional demands. Thus, knowledge is anchored in historical, economic, and political projects. Indigenous peoples have learned within colonized systems, and their knowledge is therefore no longer intact. He argued that if marginalized Indigenous peoples could delink from the colonizer's ontologies, they could create their own narratives which could support decolonization. See Alice Johnson, Chapter 4, for a description of a program that supported Indigenous children in delinking from school as it normally proceeds. Although Indigenous peoples have suffered serious cultural, social, emotional, spiritual, and physical health damage due to DWW approaches to education, in a sense, we have all been indoctrinated into DWW. Indigenous children in North America and Australia were removed from their families and taken to residential schools and government managed institutions, where they were not allowed to speak their languages, were not given sufficient food, and were physically and often sexually abused. In Australia, these children are known as the 'Stolen Generations'. Children of settler ancestry had to learn to sit still, to memorize math facts, and to study European history, and at the same time, they were taught they were the rightful rulers, and that what they were learning was their knowledge. Even though all individual children are colonized into DWW, it is very obvious who benefited from DWW, and who suffered. There is a need for critical understanding of DWW, for all people—those who suffer and those who benefit.

Through decolonization, the Other is shown not to exist ontologically, but rather is revealed to be a discursive invention. Independent thought, Mignolo (2011) believed, requires border thinking (looking to the margins of what has been taught) for the simple reason that it cannot be achieved within the categories of DWW thought and experience. Delinking and border thinking occur wherever the conditions are appropriate and the awareness of coloniality comes into being. In what ways can marginalized Indigenous peoples separate themselves from DWW ontologies? To support this move, Mignolo noted that there are two ways of knowing: the humanitas, which is ego-logical, theological, imperial, and territorial, and the anthropos, which is world-sensing, embodied in local politics, geographies, and histories. He then argued that it is through the anthropos that one can delink from the colonizers.

Ah—but here is the rub. It is so hard even to see in what ways one's being has been colonized, let alone to rid oneself of this. Mignolo would have us root our understanding in the experiential and local, but experiences are interpreted through culture. Minds (as in Vygotsky's 1978 concept of "mind in society") are colonized. As an example, Wilson (2019) wrote about the imposition of European gender binaries onto North American non-binary Indigenous peoples. DWW normalizes the practice of assigning gender at birth, in order to maintain the fiction that sex is integral to identity, and that it is a binary condition (see Kai Orca, Chapter 9, for a description of the damage that is done to children). Through "mind in society", "we all know" that babies are born as either boys or girls, but are they? Rooting our understandings in the anthropos must be associated with much critical examination.

We explored Bhabha's ideas on hybridity as another potential approach to decolonization (Bhabha, 1985, 1994; Dei, 2000; Rutherford, 1990). It is important at this point (and fitting with Bhaba) to note that we do not have to choose between two approaches: border thinking and delinking from Mignolo (2011), versus hybridity from Bhabha. Bhabha (1994) argued that our cultures are hybrids of many different influences; there is a flow of ideas and interactions that result in the complexity that is all the different components of (sub)(alternative) cultures. For example, Bhabha (1994) noted that not all Muslims agreed with the fatwa that was announced against Salman Rushdie; indeed, not only not all Muslims, but not even all Iranian Muslims of the particular sect that ordered the fatwa agreed with the fatwa. One culture is made of many different influences, and how each individual takes up those influences creates multiple hybrids. Dei (2000) wrote: "Indigenous knowledges do not 'sit in pristine fashion' outside of the effects of other knowledges ... The fact that different bodies of knowledge continually influence each other shows the dynamism of all knowledge systems" (p. 113). Individuals, as well, reveal a mix of different influences, and might draw on different ontologies and epistemologies in different situations.

Writing about the United States, Grande (2004) noted that "the quintessential dilemma of Indian peoples" was that they had to " 'pose' as domestic minorities and secure civic benefits at the price of absorption, or to claim their distinction as sovereign peoples and 'domestic dependent nations,' risking continued subjugation for cultural integrity" (p. 64). She further argued that, for United States Indigenous peoples, the tension was between maintaining their governmental systems and ontologies for working with the land, versus participating in the national democracy. Grande was describing the dichotomous choice that was offered to United States Indigenous peoples; there was no compromise possible—either give up their beliefs and ways of living to join or be left out entirely. In Grande's book, she offered various hybrid approaches United States Indigenous peoples could take. In doing so, Indigenous peoples would delink from the dominant oppressive ideologies. Delinking could involve taking up relational ways of being, recognizing the incoherence present in every category, and using one's ability to learn to keep oneself thinking and learning forever. Some of the authors of chapters in this book use the concept of delinking, within a frame of hybridity, to acknowledge the entanglements from past, present, and future colonial legacies with Indigenous ways of knowing, being, thinking, doing. All the authors in this text challenge binaries through rhizomatic thinking and doing, with no center, but rather a zigzagging network of relations.

Indigenous Knowledges

Each group of Indigenous peoples is unique because of their unique environments. In South America, an Indigenous concept is from Amerindian perspectivism (Viveiros de Castro, 1998): there is a saying that there are many natures, but only one culture, to contrast with the idea of "western multiculturalist cosmologies" (1998, p. 470). One culture (human) but many natures means that humans coproduce their culture in the context of "multinaturalism" that manifest to the specifics of their geographies, where humans are part of nature. Indigenous peoples, from every part of the world, have lived on/with the land, developing healthy relationships with the land (Dei, 2000)—otherwise, they died. With colonialism, new practices arrived and were imposed on the peoples and the land, disrupting sustainable practices. Colonialism brought differences in relationships regarding land ownership, with each of the seven countries represented in our book being different. In each country, relationships have been disrupted by colonialism, with differing approaches to colonization, and different aspects of the land exploited. In each case, the culture/nature has been disrupted.

Dei (2000) described IK at length, noting that it was "the common *good*-sense ideas and cultural knowledges of local peoples concerning the everyday realities of living" (p. 114, emphasis in original). Note that this is about everyday living. He went on to note:

> I refer specifically to the epistemic saliency of cultural traditions, values, belief systems and world views in any Indigenous society that are imparted to the younger generation by community elders. Such knowledge constitutes an 'Indigenous informed epistemology'. It is a worldview that shapes the community' s relationships with surrounding environments. (p. 114).

Note that in this part of IK, Dei is arguing the usefulness of IK for environmental attention and care. He is speaking to the concept of multinaturalism.

Lastly, Dei added, from Castellano, that there were three different aspects of IK: traditional knowledge derived from Elders, empirical knowledge derived from direct experience, and revealed knowledge from visions and intuition. Visions and intuitions, experienced by individuals, were often then shared with the community and interpreted by them, through their experiences and what they anticipated might happen. IK, Dei argued, is always personal, not universal. For others to accept this knowledge, "trust ... is tied instead to integrity, familiarity, and the perceptiveness of the 'speaker'"

(p. 114). Note the emphasis on personal relationships, leading to trust in the other.

Despite the concept of multinaturalism, despite that there is no claim to universalism and despite the belief that knowledge is always personal, there is a common aspect to IK, as described by Indigenous scholars. That is the one of relationality. In sub-Saharan Africa, the concept of M'buntu is the belief that the "I" comes into existence on entering the presence of another (Kayira, 2013; Musopole, 1994; Sindima, 1990, 1995). The greeting between people translates to "I see you, therefore I exist".

The implication from M'buntu is that humans rely on and support one another; they call one another into existence. For North American Indigenous peoples, relationality goes beyond human to human, and considers all entities, including many which DWW would consider abiotic (rocks, weather, etc.). Humans are only one species, one entity/group, amongst many, all living in more or less harmonious ways on the land. It is the relationships amongst the entities that are integral in the concept of land (Cajete, 2005; Deloria, 2001; Simpson, 2014). Deloria wrote (2001, p. 3):

> The best description of Indian metaphysics was the realization that the world, and all its possible experiences, constituted a social reality, a fabric of life in which everything had the possibility of intimate knowing relationships because, ultimately, everything was related. This world was a unified world, a far cry from the disjointed sterile and emotionless world painted by Western science.

It is this concept of relationality, whether, as in M'buntu, it is relationality to other humans, or, as in North America, to all entities, or, as in South America, to the surrounding environments—it is relationality that creates an ontology that is different from DWW. If EE is decolonized in different ways for each different context, will this decolonization rely on relationalities?

We invite you to read the chapters in this book and consider how the author's experiences and belief systems might support you as you go forward in your work to create a healthier and more just world. To support you in your understanding and choices regarding order of chapters, a brief description of the research methodologies and the chapter contents follows.

Research Methodologies and Brief Introductions to Chapters

Although as a group, we do not see the world in binaries, we are going to point out that the research methodologies we use do not fall into "quantitative approaches". There are many different kinds of quantitative research, but commonalities are the use of large sample sizes, and with the intention that research results should be replicable, apply universally, and thus be generalizable. The point of this book is that environmental (which includes social equity) issues are not generalizable but are particular for local context and nations. Therefore, qualitative research has been taken up by most authors in this book, with the exception of Kathryn Riley's postqualitative account that adopts posthumanist perspectives in Chapter 2. As such, aligned with the idea that decolonizing EE means to include multivocal approaches in the field's research, practice, and policy, this book similarly resembles a threading of different ontological, epistemological, ethical, and methodological approaches that we understand as (post)qualitative.

In the forms of (post)qualitative research in this book, each author must be aware of their hegemonic beliefs, biases, and values, and how these might affect an understanding of the research in focus. Researchers cannot "bracket out" these beliefs, biases, and values. Yet through critical awareness, the researcher can illuminate negative cases (unexpected results), and more fully acknowledge how they might be changed by the research. Lather (1986) noted that, rather than validity, reliability, and generalizability (terms used in quantitative research), other paradigms of research might expect trustworthiness and transferability. When reading a (post)qualitative study, the reader should get a sense of being there, and that the researcher's interpretation is trustworthy. The reader might not be convinced that this is the only interpretation—indeed, that is partially the point here. Given that there are other interpretations possible, authors in this book provide partial perspectives and situated knowledges, in which knowing is contingent with present moment understandings (Haraway, 1988). As well, the findings are not expected to be generalizable. Rather, (post)qualitative research provides opportunities for the reader to discern enough of the context to decide what aspects of the research might apply to their own situations and how the processes might be used in their case. Lather noted that one type of "validity" possible in naturalistic inquiry is "catalytic validity" where the researcher is changed by the research. For example, Ranjan Datta, in Chapter 8, noted how he was changed by, and how he learned from, his participants, who were knowledge keepers from the Laitu Khyeng Indigenous people at the Chittagong Hill Tracts.

In Chapter 1, Vince Anderson uses a systematic literature review and then draws on Berry's (2005) "solving for pattern" approach to identify key linkages within the literature. Vince examined the intersections between social and ecological justice issues and how theorists have integrated the two streams of literature to greater or lesser degrees.

In Chapter 2, Kathryn Riley takes up affective materiality within posthumanist perspectives to enact a non-dualistic mapping of difference between herself, as a White/Western/Settler to Canada and Buddy the Grey Wolf. Affective materiality helped Kathryn understand differences between categories, boundaries, and borders as affirmative through their relational entanglements, rather than as oppositional and dualistic (Barad, 2007; Braidotti, 2011; Lenz Taguchi & Palmer, 2013).

In Chapter 3, Marcelo Gules Borges uses a reflexive narrative research methodology from ethnographic experience (Rockwell, 2009) data to describe a teacher candidate taking up social justice issues throughout agroecology within the Brazilian science and environmental education (Andreoli & Borges, 2020; Kassiadou, 2018; Paim, 2016; Sánchez et al., 2020).

In Chapter 4, Alice Johnson utilizes a reflexive narrative methodological approach (Burawoy, 2003; Chawla & Atway, 2017; Holman-Jones et al., 2013; Lyle, 2009). Alice wrote of her experiences in a remote village, working with Indigenous students. She drew on her understandings and the research about land education (Calderon, 2014) and land-based education (Simpson, 2014, 2017), to develop a story of learning on the land with the students as a way of delinking from a colonized approach to education.

In Chapter 5, Roseann Kerr uses a critical educational ethnography informed case study (Howard & Ali, 2016), a combined methodology that explored a case of Campesino-a-Campesino in practice. This study, taking place on marginalized land with marginalized peoples in Mexico, is an example of people working together to create the understandings necessary to live sustainably with their environment.

In Chapter 6, John B. Acharibasam uses two-eyed seeing (Bartlett et al., 2012; Iwama et al., 2009; Kapryka & Dockstator, 2012), drawing from Western science and IK to contrast different views of how the world is shaped and is taught in a rural village school in Ghana.

In Chapter 7, Janet McVittie and Marcelo Gules Borges use constructivist grounded theory as their research methodology, drawing directly from the interview data, and checking back with the data as theory developed, while recognizing their beliefs cannot be "bracketed out" but must be acknowledged (Kennie & Fourie, 2014, 2015; Selden, 2005). Their chapter addresses Canadian teacher candidates of settler background coming to understand an Indigenous relational ontology while working for a sustained time on naturalized prairie.

In Chapter 8, Ranjan Datta uses a reflexive narrative methodology, similar to that of Alice Johnson, and draws on an Indigenous relational framework to consider "multiple realities, relationships, and interactions based on our traditional knowledge" (Datta, 2015, p. 107). In Ranjan's case, he describes how he learned about Indigenous beliefs about sustainability in the Chittagong Hill Tracts in Bangladesh.

In Chapter 9, Kai Orca's research is narrative methodology. As Kai commented:

> My narrative methodology involves reflection on my own life experiences cast as personal memoir, combined with literary analysis of narratives written by Two-Spirit and transgender authors. When I reflect on my life as a story, I look for patterns and meaning. I try to learn from my experiences, in order to evaluate and understand the culture of which I am a part. When I apply literary analysis to published narratives, I am reading for form as well as content, to understand how the author uses genre, figurative language, rhythm, sound, image, and theme. When I turn my narrative understandings into educational content as lesson plans, I am designing creative projects that help students see, understand and relate their own lives to the content and creative dimensions of narrative.

In the deconstruction of the gender dichotomy, Kai also deconstructs the culture-nature dichotomy.

The Epilogue resembles a relational assemblage, in which each individual author's poems are threaded together to provide a synthesis of parts into a whole. This aesthetic and lyrical method of writing is included to demonstrate how separate and discrete categories (i.e., individual authors) are relationally entangled, yet differentiated, within broader assemblages of this book.

Note

1. This is, as noted, a generalization. There were people who attempted to sound the alarm over the mistreatment of the environment prior to Carson.

References

Andreoli, V., & Borges, M. G. (2020). Apresentação do Dossiê: Pesquisas e Práticas em Educação Ambiental e Educação do Campo. *Ambiente & Educação, 25*(2), 3–18. <https://doi.org/10.14295/ambeduc.v25i2.11874>.

Andreotti, V.dO., Stein, S., Ahenakew, C., & Hunt, D. (2015). Mapping interpretations of decolonization in the context of higher education. *Decolonization: Indigeneity, Education, & Society, 4*(1), 21–40.

Barad, K. (2007). *Meeting the universe halfway: Quantum physics and the entanglement of matter and meaning.* Durham & London: Duke University Press.

Bartlett, C., Marshall, M., & Marshall, A. (2012). Two-eyed seeing and other lessons learned within a co-learning journey of bringing together Indigenous and mainstream knowledges and ways of knowing. *Journal of Environmental Studies and Sciences, 2*(4). <https://doi.org/10.1007/s13412-012-0086-8>.

Battiste, M. (2017). *Decolonizing education: Nourishing the learning spirit.* Vancouver, BC: UBC Press.

Berry, W. (2005). Solving for pattern. In M. K. Stone, Z. Barlow, & F. Capra (Eds.), *Ecological literacy: Education our children for a sustainable world* (pp. 30–40). San Francisco, CA: Sierra Club Books.

Bhaba, H. (1985). Signs taken for wonders: Questions of ambivalence and authority under a tree outside Delhi, May, 1817. *Critical Inquiry, 12*(1), 144–165.

Bhaba, H. (1994). *The location of culture.* London: Routledge.

Biedenweg, K., Monroe, M., & Wocjik, D. (2013). Foundations of environmental education. In M. Munroe & M. Kransy (Eds.), *Across the spectrum: Resources for environmental educators* (pp. 9–27). Washington, DC: North American Association for Environmental Education.

Bierhorst, J. (1994). *The way of the earth Native American and the environment.* New York William Morrow.

Blum, N., Nazir, J., Breiting, S., Goh, K. C., & Pedretti, E. (2013). Balancing the tensions and meeting the conceptual challenges of education for sustainable development and climate change. *Environmental Education Research, 19*(2), 206–217. <https://doi.org/10.1080/13504622.2013.780588>.

Bowers, C. A. (2003). "Can critical pedagogy be greened?" *Educational Studies, 34*(1), 11–21.

Braidotti, R. (2011). *Nomadic theory: The portable Rosi Braidotti.* New York & West Sussex: Colombia University.

Burawoy, M. (2003). Revisits: An outline of a theory of reflexive ethnography. *American Sociological Review, 68*(5), 645–679.

Cajete, G. (2005). American Indian epistemologies. *New Directions for Student Services,* (109), 69–78.

Calderon, D. (2014). Speaking back to Manifest Destinies: A land education-based approach to critical curriculum inquiry. *Environmental Education Research, 20*(1), 24–36.

Capra, F., & Luisi, P. L. (2014). *The systems view of life: A unifying vision.* Cambridge. UK: Cambridge University Press.

Carson, R. (1962/2002). *Silent spring: The classic that launched the environmental movement.* Boston: Houghton Mifflin, Mariner Books.

Carson, R. (1965/2017). *The sense of wonder: A celebration of nature for parents and children.* New York: Harper Collins.

Centre for Biological Diversity (2018). *The Extinction Crisis.* <https://www.biologicaldiversity.org/programs/biodiversity/elements_of_biodiversity/extinction_crisis/>.

Chawla, D., & Atay, A. (2017). Introduction: Decolonizing autoethnography. *Cultural Studies, Critical Methodologies, 18*(1), 153270861772895-8.

Datta, R. (2015). A relational theoretical framework and meanings of land, nature, and sustainability for research with Indigenous communities. *Local Environment: The International Journal of Justice and Sustainability, 20*(1), 102–113. Retrieved from: <http://dx.doi.org/10.1080/13549839.2013.818957>.

Davis, W. (2009). *The wayfinders: Why ancient wisdom matters in the modern world.* Toronto, ON: House of Anansi Press.

Dei, G. J. S. (2000). Rethinking the role of Indigenous knowledges in the academy. *International Journal of Inclusive Education, 4*(2), 11–132.

Deloria, V., Jr. (2001). American Indian metaphysics. In V. Deloria & D. Wildcat (Eds.), *Power and place: Indian education in America* (pp. 1–6). Golden, CO: Fulcrum Publishing.

Dewey, J. (1938). *Experience and education.* New York: Collier Books.

Durning, A. (1992). *Guardians of the land: Indigenous peoples and the health of the Earth* (Paper no. 112). Worldwatch Institute.

Fals Borda, O. (1980). *Science and the common people.* Yugoslavia.

Gabbidon, S. (2010). *Race, ethnicity, crime, and justice: An international dilemma.* Los Angeles, CA: Sage Publications.

González-Gaudiano, E., & Peters, M. (2008). *Environmental education: Identity, politics, and citizenship.* Rotterdam, NL: Sense Publishers.

Gough, A. (2013a). The emergence of environmental education research: A "history" of the field. In R. R. B. Stevenson, M. Brody, J. Dillon, & A. E. J. Wals (Eds.), *International handbook of research on environmental education* (pp. 13–22). New York: Routledge.

Gough, N. (2013b). Thinking globally in environmental education: A critical history. In R. R. B. Stevenson, M. Brody, J. Dillon, & A. E. J. Wals (Eds.), *International handbook of research on environmental education* (pp. 33–44). New York: Routledge.

Grande, S. (2004). *Red pedagogy: Native American social and political thought.* Toronto: Rowman & Littlefield.

Gruenewald, D. A. (2003). The best of both worlds: A critical pedagogy of place. *Educational Researcher, 32*(4), 3–12.

Haraway, D. (1988). Situated knowledges: The science question in feminism and the privilege of partial perspective. *Feminist Studies, 14*(3), 575–599.

Holman-Jones, S., Adams, T., & Ellis, C. (2013). *Handbook of autoethnography.* Walnut Creek, CA: Left Coast Press.

Howard, L. C., & Ali, A. I. (2016). (Critical) educational ethnography: Methodological premise and pedagogical objectives. *New Directions in Educational Ethnography.* Published online: 20 Dec 2016; 141–163.

Hughes, D. (1983). *American Indian ecology.* El Paso, TX: Western Press.

Hunt, S., & Holmes, C. (2015). Everyday decolonization: Living a decolonizing queer politics. *Journal of Lesbian Studies, 19*(2), 154–172.

Intergovernmental Panel on Climate Change. (2018). Global warming of 1.5°C. An IPCC Special Report on the impacts of global warming of 1.5°C above pre-industrial levels and related global greenhouse gas emission pathways, in the context of strengthening the global response to the threat of climate change, sustainable development, and efforts to eradicate poverty (V. Masson-Delmotte, P. Zhai, H. O. Pörtner, D. Roberts, J. Skea, P. R. Shukla, A. Pirani, W. Moufouma-Okia, C. Péan, R. Pidcock, S. Connors, J. B. R. Matthews, Y. Chen, X. Zhou, M. I. Gomis, E. Lonnoy, T. Maycock, M. Tignor, T. Waterfield (Eds.)).

Iwama, M., Marshall, A., Marshall, M., & Bartlett, C. (2009). Two-eyed seeing and the language of healing in community-based research. *Canadian Journal of Native Education, 32*, 3–23.

Jensen, B., & Schnack, K. (1997). The action competence approach in environmental education. *Environmental Education Research, 3*(2), 163–178.

Kapyrka, J., & Dockstator, M. (2012). Indigenous knowledges and western knowledges in environmental education: Acknowledging the tensions for the benefits of a "two-worlds" approach. *Canadian Journal of Environmental Education, 17*, 97–112.

Kassiadou, A. (2018). Educação ambiental crítica e decolonial: Reflexões a partir do pensamento decolonial latino-americano. In A. Kassiadou, C. Sánchez, D. R. Camargo, M. A. Stortti, R. N. Costa (Eds.), *Educação ambiental desde El Sur* (pp. 25–40). Macaé: Editora NUPEM.

Kayira, J. (2013). (Re)creating spaces for uMunthu: Postcolonial theory and environmental education in southern Africa. *Environmental Education Research.* <https://doi.org/10.1080/13504622.2013.860428>.

Kenny, M., & Fourie, R. (2014). Tracing the history of grounded theory methodology: From formation to fragmentation. *The Qualitative Report, 19*(Article 103), 1–9.

Kenny, M., & Fourie, R. (2015). Contrasting classic, Straussian, and constructivist grounded theory: Methodological and philosophical conflicts. *The Qualitative Report, 20*(8), 1270–1289.

Kopnina, H. (2012). Education for sustainable development (ESD): The turn away from 'environment' in environmental education? *Environmental Education Research, 18*(5), 699–717.

Lather, P. (1986). Issues of validity in openly ideological research: Between a rock and a soft place. *Interchange, 17*(4), 63–84.

Lenz Taguchi, H., & Palmer, A. (2013). A more 'livable' school? A diffractive analysis of the performative enactments of girls' ill-/well-being with(in) school environments. *Gender and Education, 25*(6), 671–687. <https://doi.org/10.1080/09540253.2013.829909>.

Lyle, E. (2009). A process of becoming: In favour of a reflexive narrative approach. *The Qualitative Report, 14*(2), 293–298. Retrieved from <http://www.nova.edu/ssss/QR/QR14-2/lyle.pdf>.

Maller, C., Townsend, M., Pryor, A., Brown, P., & St. Leger, L. (2006). Healthy nature healthy people: Contact with nature as an upstream health promotion intervention for populations. *Health Promotion International, 21*(1), 45–54.

Martusewicz, R., Edmundson J., & Lupinacci J. (2011). *Eco justice education toward democratic and sustainable communities.* New York: Routledge.

McCoy, K. (2014). Manifesting destiny: A land education analysis of settler colonialism in Jamestown, Virginia, USA. *Environmental Education Research, 20*(1), 82–97.

McKeon, M. (2012). Two-eyed seeing into environmental education: Revealing its "natural" readiness to indigenize. *Canadian Journal of Environmental Education, 17,* 131–147. <https://doi.org/10.1017/CBO9781107415324.004>.

Mignolo, W. D. (2011). Geopolitics of sensing and knowing: On (de)coloniality, border thinking and epistemic disobedience. *Confero, 1*(1), 129–150.

Musopole, A. (1994). *Being human in Africa: Toward an African Christian anthropology.* New York: Peter Lang.

Orr, D. (2010). What is education for? In D. Orr (Ed.), *Hope is an imperative: The essential David Orr* (pp. 237–245). Washington, DC: Island Press.

Paim, R. O. (2016). Educação Ambiental e agroecologia na educação do campo: Uma análise de sua relação com o entorno produtivo. *Revista Brasileira De Educação Ambiental (RevBEA), 11*(2), 240–262. <https://doi.org/10.34024/revbea.2016.v11.2152>.

Palmer, J. A. (1997). Beyond science: Global imperatives for environmental education. In P. J. Thompson (Ed.), *Environmental education for the 21st century: International and interdisciplinary perspectives* (pp. 3–12). New York: Peter Lang.

Paperson, L. (2014). A ghetto land pedagogy: An antidote for settler environmentalism. *Environmental Education Research, 20*(1), 115–130.

Piacentini, P. [Ed.] (1993). *Story Earth: Native voices on the environment.* San Francisco: Mercury House.

Roberts, H. (1998) Indigenous knowledges and Western science: Perspectives from the Pacific. In D. Hodson (Ed.), *Science and technology education and ethnicity: An Aotearoa/New Zealand perspective. Proceedings of a conference, Royal Society of New Zealand, Thorndon, Wellington, 7, 8 May 1996.* The Royal Society of New Zealand Miscellaneous Series 50.

Rockwell, E. (2009). *La experiencia etnográfica: historia y cultura en los procesos educativos.* Buenos Aires: Paidós.

Rousseau, J. J. (1762/2016). *Emile, or on education* (B. Foxley, Trans.). CreateSpace Independent Publishing Platform.

Rutherford, J. (1990). The third space: Interview with Homi Bhabha. In J. Rutherford (Ed.), *Identity: Community, culture, difference* (pp. 207–221). London: Lawrence and Wishart.

Sánchez, C., Pelacani, B., & Accioly, I. (2020). Educação ambiental: Insurgências, re-existências e esperanças. *Ensino, Saúde e Ambiente,* Número especial, 1–20. <https://doi.org/10.22409/resa2020.v0i0.a43012>.

Sauvé, L. (1996). Environmental education and sustainable development: A further appraisal. *Canadian Journal of Environmental Education, 1,* 7–34.

Schwab, J. (1963). The practical 3: Translation into curriculum. *The School Review, 81*(4), 501–522.

Seawright, G. (2014). Settler traditions of place: Making explicit the epistemological legacy of white supremacy and settler colonialism for place-based education. *Educational Studies, 50*(6), 554–572.

Seldén, L. (2005). On grounded theory – With some malice. *Journal of Documentation, 61*(1), 114–129.

Simpson, L. B. (2011). *Dancing on our Turtle's back: Stories of Nishnaabeg recreation, resurgence, and a new emergence.* Winnipeg, MB: Arbeiter Ring Publishing.

Simpson, L. B. (2014). Land as pedagogy: Nishnaabeg intelligence and rebellious transformation. *Decolonization: Indigeneity, Education & Society, 3*(3), 1–25.

Simpson, L. B. (2017). *As we have always done: Indigenous freedom through radical resistance.* Minneapolis, MN: University of Minnesota Press.

Sindima, H. (1990). Liberalism and African culture. *Journal of Black Studies, 21*(2), 90–209.

Sindima, H. (1995). *Africa's agenda: The legacy of liberalism and colonialism in the crisis of African values.* Westport, CT: Greenwood Press.

Stapp, W. (1969). The concept of environmental education. *Journal of Environmental Education, 1*(1), 30–31.

Tilbury, D. (1995). Environmental education for sustainability: Defining the new focus of environmental education in the 1990's. *Environmental Research, 1*(2),1995–212. <https://doi.org/10.1080/1350462950010206>.

Tuck, E., McKenzie, M., & McCoy, K. (2014). Land education: Indigenous, postcolonial, and decolonizing perspectives on place and environmental education research. *Environmental Education Research, 20*(1), 1–23.

Tuck, E., & Yang, K. W. (2012). Decolonization is not a metaphor. *Decolonization: Indigeneity, Education & Society, 1*(1), 1–40.

UNESCO/UNEP. (1975). *The Belgrade charter: A framework for environmental education.* <https://www.eusteps.eu/wp-content/uploads/2020/12/Belgrade-Char ter.pdf>.

UNESCO/UNEP. (1977). *Final report. Intergovernmental conference on environmental education,* organized by UNESCO in cooperation with UNEP. Tblisi, USSR, 14–16 October, 1977. <https://www.gdrc.org/uem/ee/EE-Tbilisi_1977.pdf>.

UNESCO. (1987). *Report of the World Commission on Environment and Development: Our common future.* <https://sustainabledevelopment.un.org/content/documents/5987our-common-future.pdf>.

Viveiros de Castro, E. (1998). Cosmological deixis and Amerindian perspectivism. *The Journal of the Royal Anthropological Institute, 4*(3), 469–488.

Vygotsky, L. S. (1978). *Mind in society: The development of higher psychological processes* (M. Cole, V. John-Steiner, S. Scribner, & E. Suberman, Eds.). Harvard University Press.

Wals, A. E. J., & Jickling, B. (2000). Process-based environmental education: Seeking standards without standardizing. In B. B. Jensen, K. Schnack, & V. Simovska

(Eds.), *Critical environmental and health education: Research issues and challenges* (pp. 127–148). Copenhagen: The Danish University of Education, Research Centre for Environmental and Health Education.

Wilson, A. (2019). Skirting the issues: Indigenous myths, misses, and misogyny. In K. Anderson, M. Campbell, & C. Belcourt (Eds.), *Keetsahnak / our murdered and missing Indigenous sisters.* (pp. 161–174). Edmonton, Alberta: University of Alberta Press.

World Wildlife Federation (2018). How Many Species are We Losing. <http://wwf.panda.org/our_work/biodiversity/biodiversity/>.

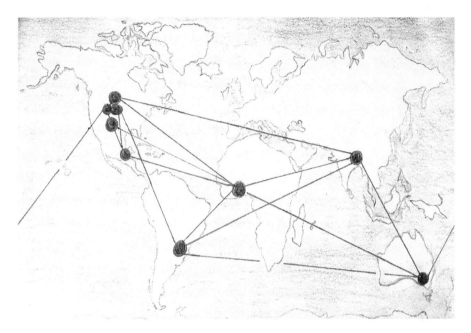

"A Global Assemblage" (2021)
Zuzana Morog

1. Integrating Social and Ecological Justice Inquiry

Vince Anderson

My family and I make our home on Treaty Six Territory and the Homeland of the Métis. I pay respect to the First Nations and Métis ancestors of this place and honour our relationship with one another. I wish to reaffirm my dedication to actively working towards decolonization and the transformation of all other forms of oppression as I experience the privilege of being a part of this project.

The objective of this edited work, *Decolonizing Environmental Education for Different Contexts and Nations*, is to apply the lens of decolonization to the theory and practice of environmental education (EE). This requires the integration of two fields of inquiry, decolonization and environmental education, each containing a great deal of variation and complexity. The integration of social and environmental (or ecological)[1] fields of inquiry to better understand current crises that impact the world is becoming increasingly common (Capra, 2005; Greenwood, 2003). In this chapter, I explore a series of interconnected elements that are found within social justice and ecological justice literatures. The purpose of this exploration is to further discussion on what it means to conduct inquiry at the intersections of social and ecological justice. In this book, the social and ecological are theorized as interconnected and interdependent dimensions of necessary teaching and learning. This chapter supports the overarching goal of the book by outlining one approach to integrating social and ecological justice theories. In the chapter, I identify key theoretical elements found within a diverse set of social and ecological justice literatures and consider how these elements might be combined in the development of pedagogies.

The fields of inquiry I draw from in the discussion include, critical theory, critical race theory, feminist and queer theories, anti-oppressive theory, decolonization, environmental justice, environmental racism, climate justice, critical ecopedagogy, and Land education. For each field of inquiry addressed in the chapter, my review of the theory is focused and purposive. Each field addressed herein is varied and complex, just as the fields of decolonization and EE are. Therefore, a comprehensive review and representation of the fields is impractical. My goal is to explore potentially constructive theoretical perspectives that might inform the development of integrated social and ecological justice pedagogies.

An increasing number of researchers, educators, and advocates working toward a wide array of justice objectives emphasize a need for a clearer engagement with social and ecological aspects of problems observed in the world. What is increasingly seen as going wrong involves an integrated matrix of *social oppression* and *ecological degradation*. Scholars and citizens have begun centering analysis on this interconnection of the social and ecological as a foundational basis for exploring humanity's past, present, and future (Calderon, 2014; Greenwood, 2003; Kahn, 2010; Pulido, 2015; Tuck, McKenzie, & McCoy, 2014).

In this chapter, the *social* includes the interpersonal, institutional, structural, and political dimensions of human experience, wherein values, customs, ideologies, discourses, and power play out in complex ways that are suffused throughout our collective past, present, and future (Bourdieu, 1984; Butler, 1990). Social oppression (or simply oppression) is understood as the lived consequences of a history of unequal power relations among social groups, where the dominant group controls the material, structural, and discursive aspects of social reality, employing various means to preserve and enhance its domination (Kumashiro, 2000, 2009; Sensoy & DiAngelo, 2012). The hierarchy, objectification, and violence deployed through this unequal power relation is infused in all socio-political institutions, and is hegemonic when the discourse flowing from the dominant group becomes the default, taken-for-granted perspective within the society as a whole (Butler, 1990; Kumashiro, 2000).

The *ecological* is interpreted as the interconnected relationships among networks of living systems that establish the foundations of life. As Fritjof Capra (2005) explains,

> First, *every* living organism, from the smallest bacterium to all the varieties of plants and animals, including humans, is a living system. Second, *the parts of living systems* are themselves living systems. A leaf is a living system. A muscle is a living system. Every cell in our bodies is a living system. Third, *communities of organisms*, including both ecosystems and human social systems such as families, schools, and other human communities, are living systems. (p. 19, Italics in original)

Ecological degradation is understood as the state and/or process in which the attributes of living systems are disrupted, degraded, and/or pushed beyond the limits of recovery. Ecological degradation encompasses disruptions in living systems at multiple levels of organization, from an individual stream or aquifer to global climate patterns (Capra, 2005; Greenwood, 2009).

Exploring the 'Social' in Social and Ecological Justice Theories

In order to develop an idea of how to engage the *social* and *ecological* in justice theorizing, it seems productive to explore current articulations of both social justice and ecological justice, as well as ideas of how theory might be mobilized towards effecting change. It is commonplace for writing on social justice to provide some form of definition. In order to develop a definition, concepts such as equity, dignity, human rights, and freedom from discrimination and violence are often emphasized as responses to historical, institutional, political, and economic dimensions of oppression (Dimick, 2012; Rose & Cachelin, 2014). In addition, researchers often emphasize the necessity to critically examine ways social group identifiers—race, class, gender identity, sexual orientation, ability, age, nationality, religion, body type, etc.—are related to multiple dimensions of historical and contemporary oppression (Warren et al., 2014).

In their book, *Is Everyone Really Equal?*, Ozlem Sensoy and Robin DiAngelo (2012) critique the notion of social justice as it tends to be interpreted by the public, and how it is interpreted within some spheres of research and academic practice. The authors propose that while a majority of individuals in a liberal pluralist society would assert their support for principles of equal opportunity and protection from discrimination and violence, there are limits to what the same individuals are willing to question and what they are willing to advocate for in terms of social change. The authors describe these limitations as a form of "social justice illiteracy" (p. xvii), or an inability to accurately recognize and interpret the historical and contemporary factors that give rise to inequality, discrimination, and violence. Albert Memmi (2000), in his book *Racism*, begins with a similar assessment, "There is a strange kind of tragic enigma associated with the problem of racism. No one, or almost no one, wishes to see themselves as racist; still, racism persists, real and tenacious" (p. 3). For these researchers, a key consideration for defining social justice involves being aware of a limitation in the way the term is understood and used both within popular discourse, and in some approaches to research and advocacy.

Throughout their book, Sensoy and DiAngelo (2012) develop their description of social justice by contrasting it with populist conceptualizations they examine in everyday discourses. They assert that in addition to

challenging the obvious and overt injustices that tend to inspire public out-
rage, social justice must begin with a keen awareness that everyday (seemingly mundane) practices are always and already influenced by unequal power
relations among social groups—mediated by, for instance, race, class, and/
or gender identity. The unequal power relations have a deep and profound
history, and are reproduced through material and discursive systems, per-
petuated in part as a result of social justice illiteracy. Finally, the authors
contend that in order for one to have a meaningful connection to social justice, there must be an examination of one's own social group membership.
They argue that such an examination is a key and necessary step in research,
educational programming, policy writing, advocacy, and so on (Sensoy &
DiAngelo, 2012).

As researchers move from general ideas of equity and rights toward more
specific justice issues, they typically draw on a complementary array of fields
and schools of thought, including but not limited to, critical theory, critical race theory, feminist theory, queer theory, anti-oppressive theory, and
decolonization. The brief overview of each provided herein is helpful for
the purposes of this chapter, but to gain a more in-depth understanding of
what is encompassed within each area, a deeper exploration is required. In
literature on social justice, researchers will often transition back and forth
between more general conceptualizations of social justice and more focused
conceptualizations articulated within specific research areas. Social justice is
often being conceptualized in relation to one or several of the more focused
research areas found under this umbrella term. This is significant to consider
when looking for the connections and patterns present within diverse social
justice theorizing.

The discussion below is focused and selective. The breadth of the body of
work developed around each area necessitates a selective reading. The research
areas and pieces selected within those areas in no way represent an exhaustive
sample of relevant literature. The aim of this section is to strategically identify
a number of potentially foundational elements that can support a more holistic, integrated, and serviceable approach to social justice theorizing.

Insights from Critical Theory

According to Henry Giroux (2009) (see also McLaren & Kincheloe, 2007),
critical theory grew out of a collaborative body of work in the mid-twentieth
century by a group of theorists known as the Frankfurt School (e.g., Adorno,
Horkheimer, Marcuse, Habermas). In general, the aim of the researchers
was to illuminate and confront ways in which laws, institutions, policies,

scholarship, media, values, cultural practices, and so on, favor, empower, and enrich some members of the society, while silencing, disenfranchising, and oppressing others. According to Giroux (2009), the School's "members developed a dialectical framework by which to understand the mediations that link the institutions and activities of everyday life with the logic and commanding forces that shape larger social totality" (p. 27). Thus, critical theory guides one to deconstruct structures and systems in order to reveal how they really impact individuals within the society. Frequently, policies and procedures that have a profound effect on the day-to-day lives of individuals are created and recreated with little serious attention. Critical theory brings to bear questions such as, who designs the policies and procedures? Who benefits? Who suffers? What assumptions are being made? And, what voices and perspectives are missing? While critical theory may be concerned with illuminating inequities related to all social groups, it often focuses on means and control of production and wealth, and the ways these arrangements influence legislation and governance. In other words, the focus is often on the political economy, which is read through a class analysis, and therefore has connections with Marxist and Neo-Marxist theory (Giroux, 2009).

According to Giroux (2009), foundational ideas guiding early work in critical theory included, self-conscious critique, social transformation, and emancipation. These ideas organized research in a way that departed from the instrumental rationalism that was common in the Academy, as well as other influential institutions. Giroux argues that, through centring these commitments, there was a shift in both the object of study and the approach to research. For example, "By examining notions such as money, consumption, distribution, and production, it becomes clear that none of these represents an objective thing or fact, but rather all are historically contingent contexts mediated by relationships of domination and subordination" (p. 27). Critical theory recognizes that the institutions and systems shaping everyday life are not neutral entities, emerging naturally out of a free and democratic society. Rather, they are designed and produced within a historical moment that is hierarchical and therefore (re)produces oppression. This recognition is foundational to social justice theorizing, and is key to confronting all forms of injustice and oppression (Giroux, 2009).

Insights from Critical Race Theory

The research area known as critical race theory began with the focused and necessary objective to debunk the "science of race" of the nineteenth and early twentieth centuries (Memmi, 2000). The linked ideas of pure race and hierarchies of race were long held as truths in many centers of power and knowledge creation. As Albert Memmi (2000) outlines, given the unintelligible premises on which these theories were based, it required a studied myopia and considerable political commitment to maintain this supposedly scientific field of inquiry. Although the historical and biological foundations behind the hierarchy of pure races have been rejected for the most part, the phenomena of race and racism remain pervasive mediators of daily social experience. Thus, critical race theory has evolved—just as racism itself has evolved—to interrogate the presence of racism in contemporary liberal discourse (Memmi, 2000). Critical race theory recognizes that the phenotypical characteristics used to distinguish race (skin color, hair texture, facial bone structure, etc.) are always and already marked with meaning and value, and influence both identities and understandings of the Other (Ladson-Billings, 2009; Sensoy & DiAngelo, 2012). These meanings and values are not derived from personal experience, nor are they individual or independent of historical and contemporary systems of power; rather, race discourses are passed down, consumed, repeated, and internalized, and exert influence on and through both the dominant and minoritized groups (hooks, 1994; Leonardo, 2009; Ladson-Billings, 2009; LaRocque, 2010; Memmi, 2000).

In his book *Race, Whiteness, and Education*, Zeus Leonardo (2009) explores the most effective way to promote understandings of critical race theory, taking into consideration both various perspectives (or camps) within the field and the state of public (mis)understandings around race and racism. Leonardo concludes that the current challenge and potential of critical race theory lies foremost in the exploration and analysis of whiteness. Leonardo's vision is to engage the "problem of whiteness and white supremacy within the color-blind era", toward bringing to light "the codes of white culture, worldview of the white imaginary, and assumptions of the invisible marker that depends on the racial other for its identity" (p. 9). The objective of illuminating these codes, worldviews, and assumptions is to name and counter "direct processes that secure [white] domination and the privileges associated with it" (p. 9). While the study of race and racism often focuses on the discrimination and violence that is committed against people of color, Leonardo (2009) recognizes that it is at least as important to illuminate and interrogate the way whiteness as an identity category is produced and defended by individuals and structures of power.

Another key element of critical race theory, outlined by Gloria Ladson-Billings and William Tate (2006), is the importance of story, the sharing experiences of racism as a valid and authoritative element of research. For Ladson-Billings and Tate (2006), story in research validates the lived, nuanced ways racism shapes lived reality; it can serve as a means of psychic preservation against the internalization of oppression; and, finally, "naming one's own reality with stories can affect the oppressor" (p. 21). In order to understand race and racism as pervasive social phenomena, critical race theorists utilize tools familiar to many social scientists, including analyses of history, demographics, legislation, political-economy, and so on. But, there is also an understanding that these tools cannot illuminate everything about how race and racism play out in day-to-day experience. In critical race theory, story is used as a link between general ideas and particular circumstances.

Insights from Feminist and Queer Theories

In *Gender Trouble*, Judith Butler (1990) opens her discussion by exploring a question of how the subject in feminism is interpreted and addressed. It is clear to Butler that feminism's objective "seeks to extend visibility and legitimacy to women as political subjects" (p. 3), and to confront and transform the patriarchy and phallocentrism common within society. What concerns Butler, and has invigorated debate within the field, is how the principal subject of feminism is represented by researchers. Butler states, "The very subject of women is no longer understood in stable or abiding terms. There is a great deal of material that not only questions the viability of 'the subject' as the ultimate candidate for representation, or indeed, liberation, but there is very little agreement after all on what it is that constitutes, or ought to constitute, the category of women" (p. 4). For Butler, it is undeniable that there is a force of domination and oppression that extends from categories of male/masculinity/heterosexuality. However, the way in which this force plays out on a daily basis in the lives of diverse individuals who do not have membership within the dominant group is complex, and does not conform to reductive, binary conceptualizations of "the subject." Butler (1990) further complicates the subject of feminist (and/or social justice) research and advocacy by stating: "If one 'is' a woman, that is surely not all one is: the term fails to be exhaustive ... because gender is not always constituted coherently or consistently in different historical contexts, and because gender intersects with racial, class, ethnic, sexual, and regional modalities of discursively constituted identities" (p. 6).

The problematizing and reframing of the subject in feminism creates new opportunities for understanding the nature of the oppression originating from the dominant center, and a vision for what social justice might look like in practice. A key consideration for Butler (1990) in her discussion of gender as a category of study is the function of intelligibility, and how this is entangled with performativity and the representational politics of sex, gender, and desire. For Butler, the absence of intelligibility—to be recognized, understood, and represented—exists as a crucial factor in the way oppression operates as a daily reality. Arguing the importance of opening the possibility of advocacy in relation to the intelligibility of the subject, Butler asserts, "One might wonder what use 'opening up possibilities' finally is, but no one who has understood what it is to live in the social world as what is 'impossible,' illegible, unrealizable, unreal, and illegitimate is likely to pose that question" (p. viii). Thus, any approach to research or advocacy must be critical of the way the research or advocacy approach itself may omit, deny, and/or negate those whom the effort aims and/or purports to represent. Butler's (1990) contribution to troubling the categories of sex, gender, and desire blends into a discussion of queer theory. Queer theory similarly seeks to illuminate and confront oppression extending from male-masculine-heterosexual dominance, but also endeavors to center non-heterosexual and non-cisgender realities as the lenses through which analysis is conducted (Kumashiro, 2000).

Max Kirsch (2000) begins his discussion of queer theory by establishing its foundation in postmodern and post-structural theoretical traditions. The significance of these movements to Queer theory, according to Kirsch, is an interrogation of knowledge constituted within the context of power and dominant discourses, and which then serves to reproduce regimes and systems of power. Kirsch continues, "What demarcates Queer theory from its postmodernist and post-structuralist foundations is its referral to a range of work 'that seeks to place the question of sexuality as the centre of concern, and as the key category through which other social, political, and cultural phenomena are to be understood'" (Edger & Sedgewick, cited in Kirsch, 2000, p. 33). Although Queer Theory centers analysis on gender and sexuality, it involves a method of critique that is relevant to all modes of interrogating unequal power and oppression. Kirsch references how the New York advocacy group Queer Nation early embraced the term *queer* "to signify a free-flowing organization of resistance that promises to transcend mainstream politics and include all who were against any set conceptions of gender, sexuality, and power" (p. 33). Kirsch discusses the "principle" of queer as "the disassembling of common beliefs about gender and sexuality from their representation in film, literature, and music to their placement in the social and physical sciences," and the "activity of 'queer' [as] the 'queering' of culture, ranging from the reinterpretation of characters involved in cinema to the deconstruction of historical analyses" (p. 33). Thus, the notion queer signals a departure from normative, taken-for-granted codes and protocols of identity politics and performativity (Butler, 1990), and a challenge to the assumed fixidity and stability of social group categories.

Kevin Kumashiro (2009) outlines a similar critique of "normal" as he contemplates forms of activism to which he is drawn: "Being normal requires thinking in only certain ways, feeling only certain things, and doing only certain things. And it punishes those who do not conform, such as those who do not look normal, or love the right kind of person, or value the important things" (p. 52). From this perspective, the question of oppression extends beyond instances and events of discrimination and violence—though these are never omitted from analysis—to the ways we are all embedded and entangled in hierarchies set up by the norm. Although all are influenced by the hegemony and reproduction of the norm, each individual negotiates the landscape of privileging and marginalizing uniquely, depending on which sources of privilege one has access to. Such investments pose a significant challenge to implementing change aligned with social justice principles. For Kumashiro (2009), investment in the normal is linked with the way compliance to the norm offers comfort and opportunity in both political and psychosocial ways.

Insights from Anti-Oppressive Theory

In the opening of Kevin Kumashiro's (2000) article, *Toward a Theory of Anti-Oppressive Education*, he offers a point of reference for engaging the theory: "In an attempt to address the myriad ways in which racism, classism, sexism, heterosexism, and other forms of oppression play out in schools, educators and educational researchers have engaged in two types of projects: understanding the dynamics of oppression and articulating ways to work against it" (p. 25). Reflected in this introductory statement is Kumashiro's (2000) assertion that there is value in examining ways the phenomena of racism, classism, sexism, and heterosexism intersect and interrelate in daily experience. The approach balances a goal of understanding the dynamics of each form of oppression with a recognition that within lived experiences multiple forms of oppression play out simultaneously. Thus, for Kumashiro (2000), anti-oppressive theory originates in the moments and spaces where multiple dimensions of oppression shape how difference and inequality impact day-to-day life.

Kumashiro (2000) describes oppression as "a situation or dynamic in which certain ways of being (e.g., having certain identities) are privileged in society while others are marginalized" (p. 25). Although this definition is quite general, it focuses research and advocacy on processes of privileging and marginalizing, rather than on a particular form of oppression. This is not to say that an anti-oppressive approach avoids naming realities of racism, classism, heterosexism, etc.; in fact, naming and exposing these are core objectives. Anti-oppressive theory explores social group identifiers not as individual subjects of study, but as integrated and interconnected mediators of social organization and experience that promote the hierarchies that ensure the privilege of dominant groups. In other words, anti-oppressive theory examines the intersection of race, class, gender, sexual orientation, nationality, and ability and is always working on and through social actors in ways that privilege and marginalize. In this project, anti-oppressive theory draws on critical race, queer, and feminist theories in order to develop a more coherent understanding of the intersectionality of multiple forms of oppression (Kumashiro, 2000).

Insights from Decolonization

In order to address current realities connected to a lack of social justice, it is necessary to examine the role that colonialism has played (and continues to play) in the production of unequal power, and the endemic discrimination and violence that are involved. In *The Wretched of The Earth*, Franz Fanon (1963) describes how colonizers dehumanize and objectify the Peoples and Lands over which they seek domination. In order to justify the violence required to achieve domination, Fanon outlines the colonial project in terms of the "fabrication" of the colonial subject, which becomes inscribed in the colonizer's law, philosophy, science, and worldview. The colonizer's identity as well is constructed in opposition to this fabrication. Fanon (1963) discusses colonialism's project of domination as an historic event that has profound implications across generations and that continues to shape contemporary realities.

In her book, *When the Other is Me*, Emma LaRocque (2010) reviews decades of literature used in the fabrication of the colonial subject in Canada (though her discussion is relevant across the globe). LaRocque outlines her findings in terms of a binary, established as the "civilized" versus the "savage" (or civ/sav):

> [T]he civ/sav dichotomy is spelled out in terms of cultural 'traits' that reflect binary opposites, each civilized trait corresponding, inversely, with a savage one. In Canadian terms, civilization is consistently associated with settlement, private property, cultivation of land and intellect, industry, monotheism, literacy, coded law and order, Judeo-Christian morality, and metal-based technology. Civilization stands for what is illuminated, progressive, and decent, while savagery is its shadowy underside. Such a 'civilization' is repeatedly outlined against

'Indian savagery,' in which savagism is seen as a psychosocial fixed condition, the antithesis of the highest human condition. Indians, then, by contrast, are delineated as wild, nomadic, warlike, uncultivating and uncultivated, aimless, superstitious, disorganized, illiterate, immoral, and technologically backwards (p. 41).

Aman Sium, Chandni Desai, and Eric Ritskes (2012) describe contemporary modes of colonialism as a reproduction of the civ/sav discourse, which reinforces a sense of entitlement to colonized Lands (manifest destiny) that continues to shape colonial practices and justify the ongoing erasure of Indigenous knowledges and realities. Delores Calderon (2014) describes decolonization as "uncovering how settler colonial projects are maintained and reproduced" (p. 28). For Calderon, a significant obstacle to decolonization is settler territoriality, which represents settler control over Land, resources, and the institutions that control these, but also the perceived entitlement to this control and the agency to dictate the future direction of Indigenous-settler relations (see also Tuck, McKenzie, & McCoy, 2014). Calderon describes a co-optation by settlers of Indigenous culture, identity, and epistemology, and positions this cooptation as linked to a deeper colonial process in which the colonizer sets the terms for how relations can be negotiated—even within justice-based priorities such as decolonization and reconciliation. This is consistent with Fanon's (1963) description of the colonial process as an historical event with long-term, ongoing significance. Decolonization, therefore, necessitates an examination of how colonialism and settler territoriality are understood and represented, and how discourses associated with Land and Indigenous-settler relations are delimited by a history of fabrication and cooptation that cannot be separated from present realities.

Initial Connections Identified Across the Theories

The goal of the above discussion has been to examine a broad range of considerations for social justice research and advocacy, and to see if some form of operational framework might emerge. One significant outcome of this type of review is to reveal how multiple approaches to social justice overlap and complement one another, creating a more complete and coherent picture of how oppression operates within lived realities. The review suggests that social justice research will be enhanced when it starts by acknowledging that socio-political structures are infused with unequal power relations that play out in complex ways. The dynamics of unequal power are manifested across a multi-variable set of social group categories (Giroux, 2009; Kumashiro, 2000; Sensoy & DiAngelo, 2012). The dominant group, as in the examples of racism and colonialism, employs all available means (juridical, scientific, political, narrative) to "fabricate" the Other in order to justify and reproduce oppressive and violent structures (Butler, 1990; Fanon, 1963; LaRocque, 2010). This fabrication is established to dehumanize and to justify exploitation, but is also used to define the dominant group in opposition to the negative attributes imputed to the targeted subaltern group. The dominant group, as in Leonardo's (2009) discussion of whiteness, becomes invested in the identity established in opposition to its derogatory fabrication of the Othered subject. Thus, social justice research must work towards both transforming discrimination and violence, and challenge the "codes," "worldview," and "assumptions" of the dominant group (Leonardo, 2009).

Another consequence of this fabrication is it serves as the criteria for defining the subject as intelligible in the eyes of the dominant group (Butler, 1990). Butler (1990) argues that the performativity of expectations for what it means to be sexed, gendered, and desiring subjects, positions difference or "queerness" as unknowable and thus forced clumsily into partial and largely inaccurate stereotypes. Those who hold such stereotypes subsequently enact the partial and inaccurate views with some combination of uncertainty, distrust, disdain, and ultimately, prejudice and violence. Thus, intelligibility becomes an important guiding consideration for learning and advocacy. Critically analyzing the way intelligibility plays a role in the power dynamics of daily interactions provides a constructive basis for establishing cross-disciplinary processes for inquiry and transformation. These initial connections inform a process of inquiry that may be applied as one revisits diverse theories within social and ecological justice.

Engaging the "Ecological" in Social and Ecological Justice

The array of ecological justice research is as large as the body of work focusing on social justice. For the purpose of this paper, discussion will concentrate on orientations to research that focus on the interrelationship between ecological and social, political, and economic systems. Specifically, the review centers on analyses of these relationships employing critical perspectives that are consistent with those that guided the previous section. Cognizant of the profound need to confront and transform oppression, this section explores how ecological and social justice research may be engaged in concert in ways that enhance the cogency of both. In parallel with the format already used for reviewing social justice literature above, the following section provides a sample of ideas explored within a number of complementary research areas. The selected fields of inquiry include environmental justice, environmental racism, climate justice, critical ecopedagogy, and Land education. These fields of research all emphasize the intersection between social and ecological systems and bring to bear a critical examination of ecological problems.

In 1983, the World Commission on Environment and Development (WECD), or Brundtland Commission, was formed by the United Nations to outline a framework for global human activity aimed at preserving the ecological stability of the Earth. After four years of research and consultation, the Commission released its report entitled *Our Common Future* (or The Brundtland Report) (United Nations, 1987). The report was significant because it would influence a collection of rapidly emerging research areas related to environment, ecology, and sustainability (Nolet, 2016). The Report emphasized that sustainability must involve a rigorous engagement with issues of poverty, growth, resource use, food security, water availability, population, pollution, and armed conflict (United Nations, 1987).

For researchers concerned with these fundamental issues, it quickly became clear that their efforts must involve addressing questions that go well beyond environmental preservation. Indeed, today it is impossible for many to think of the challenges of poverty, resource use, pollution, food and water availability, population growth, and armed conflict as being independent of classism, racism, sexism, heterosexism, and colonialism. In 2015, the UN General Assembly released an updated list of Sustainable Development Goals entitled *17 Goals to Transform Our World* (United Nations, 2015). Many of the key issues are reiterated from the 1983 report, however, the document includes specific language related to the transformation of injustice and oppression that exists globally in the forms of classism, racism, and sexism—though unfortunately issues of heterosexism and colonialism are not explicitly addressed.

Insights from Environmental Justice

Environmental Justice operates as a subset of a broader field of environmental (and/or ecological) research and action. Early orientations of environmental research concentrated on goals of conserving nature spaces, protecting wildlife, mitigating polluting industrial and household activities, and promoting a deep relationship with nature (Palmer, 1998). While these priorities undoubtedly remain essential for many environmental researchers, some have also moved to engage with deeper questions of human interconnection with the environment. This emerged in no small part due to a series of international conferences and commissions on the global environment beginning in 1967—one of which being the aforementioned WCED (Palmer, 1998). As seen in *Our Common Future* (United Nations, 1987), environmental researchers have long held an implicit understanding that measures to safeguard the stability and resilience of the planet's natural systems can only be achieved in relation to and together with social justice.

In 1987, Reverend Benjamin Chavis and the United Church of Christ's Commission for Racial Justice (UCC) undertook a research project focusing on how the benefits and harms of environmental destruction are distributed disproportionately along class lines and, in particular, race (UCC, 1987). This study was inspired by a series of actions taken across the United States by predominantly Black, Latino/a, and Indigenous communities to resist plans to establish toxic and nuclear waste sites adjacent to their homes (Gosine & Teelucksingh, 2008). Inspired by these community actions, the study headed by Reverend Chavis, along with a number of complementary studies (Bryant & Mohai, 1992; Bullard, 1990), enabled environmental researchers and activists to establish unequivocally the correlation between environmental destruction and classism and racism. This development in the field solidified that the conditions at the core of environmental destruction have a clear and important relationship with the conditions at the core of social oppression and violence. One clear implication is the way *bodies* are devalued based on race- and class-based codes, and therefore systemically "zoned" as bodies suitable for contamination.

Another important parallel between questions of social justice and ecological justice is the way that *Land* is understood, represented, and engaged. From the perspective of Reverend Chavis (UCC, 1987), Robert Bullard (1990) and other environmental justice researchers (Bryant & Mohai, 1992), the key consideration is how Land (also water and air) is determined appropriate for contamination. At the heart of the critique are patterns of deep-seated discrimination that are embedded in legal, economic, and ideological orientations to Land. As such, the focus of research centers on historical and contemporary discriminatory housing policies, roots of economic disparity, inconsistent environmental regulations, representation in government, public indifference to the poisoning of bodies that are devalued, and so on (Gosine & Teelucksingh, 2008). While these are all essential, as we will see below, researchers interested in social and ecological justice engage *Land* in additional significant ways.

Research that focuses on decolonization strives to expose and confront how historical and contemporary narratives and structures have and continue to empower this appropriation and exploitation of Land. Contemporary forms of colonization—whether in the forms of settler colonialism, military-backed imperialism, or economic globalization—are largely driven by the interests of those in positions of power to control and draw wealth from Land and the resources available therein. This includes the exploitation of cheap labor and operating in the virtual absence of environmental regulations.

In Canada, many of the most contentious struggles around First Nations, Métis, and Inuit rights center around Land and resources. Invariably, the settler colonial objective seeks control of Land and the ability to draw wealth from it (Gosine & Teelucksingh, 2008). Given this position, it is easy to see that narratives grounded in racism and classism serve a function in the contemporary colonial project, and therefore that decolonization be taken up as a key focus of environmental justice. This can also be seen globally in the forms of armed conflict and/or geopolitics employed to gain access to commodities for influential corporations, and in corporate purchase of or license over foreign Land for mining, agriculture, waste disposal, etc., all at the expense of local populations. Narratives and structures grounded in racism and classism serve these arrangements as well, both as moral justification and as a means to gain support (or avoid resistance) from leaders and settlers on colonized Lands.

Insights from Environmental Racism

Closely related to Environmental Justice, the more narrowly focused purview Environmental Racism, was initially based on the findings of Reverend Chavis and others (Bryant & Mohai, 1992; Bullard, 1990) that race is the strongest indicator of where landfills, waste sites, incinerators, and polluting industries are located in the United States (UCC, 1987). Andil Gosine and Cheryl Teelucksingh (2008) demonstrate in their book *Environmental Racism in Canada* that this reality is present also in Canada and elsewhere. This finding has required environmental justice researchers to look more deeply into the role that racism plays in the unequal distribution of environmental contamination, and therefore, to pursue a closer understanding of how racism operates as a mediator of day-to-day lived reality.

While it may be fairly straightforward to show empirically the correlation between the zoning of polluting industries and the locations of communities of color, as Laura Pulido (2015) discusses, it is much more difficult to show how race and racism factor in all socio-economic policies and practices that give rise to this correlation. As Pulido (2000) delves into an analysis of why it is that communities of color bear a far greater burden of environmental contamination than those inhabited predominantly by whites, she concludes that researchers must first understand that all spaces are circumscribed by race and racism, and that these play out in multiple and nuanced ways. This perspective aligns closely with the discussion of critical race theory included above.

Pulido has argued (2000, 2015) that research on environmental racism must go beyond popular understandings of racism, engaging analyses of white privilege and white supremacy to gain more meaningful understandings. In order to clarify what she sees as missing within popular understandings, Pulido (2000) contrasts assumptions of racism as deliberate, overt, and malicious acts committed by individuals with an understanding that "racial meanings are embedded in our language, psyche, and social structures" and that "[t]hese racial meanings are both constitutive of racial hierarchies and informed by them" (p. 13). Consistent with critical race theory, Pulido argues that the goal of environmental racism research is to interrogate the white-ordered historical and contemporary policies that have created "sociospatial processes of inequality" that are at the heart of environmental racism (p. 14). Pulido does not contend that deliberate, malicious acts of racism and environmental racism do not exist; rather, her objective is to emphasize that such acts are a part of a deeper and more profound phenomenon that impacts social (and sociospatial) reality (Pulido, 2000, 2015).

Insights from Climate Justice

As may be inferred, Climate Justice research examines the way consequences of climate change are largely borne by the most vulnerable populations across the globe (Roser et al., 2015; Wallimann-Helmer, 2015). Studies in climate science are inevitably complex given the intricacy and interdependency of systems that must be understood to gain a sense of the potential harm being done to global ecological stability (Knutti & Rogelj, 2015; Nightingale et al., 2020). While the subject is inherently complex, understanding the dynamics involved when certain individuals make decisions from within geographic centers of power (Peet, 2007) that have a profound and damaging effect on social and ecological systems of those who are already marginalized adds an entire additional layer of complexity. Dominic Roser, Christian Huggel, Markus Ohndorf, and Ivo Wallimann-Helmer (2015) assert that recent trends in climate science have been driven by findings from both natural and social science researchers that emphasize questions of justice when investigating the consequences of climate change. The authors add that this trend has emerged in conjunction with what they see as a growing tendency of political philosophers placing an emphasis on the imminent threat climate change poses as core imperative for contemporary philosophy (p. 350).

Similar to environmental justice, climate justice is concerned with the unequal distribution of environmental exploitation and destruction, and how this is mediated by the intersection of uneven social power, valuations of

Land (also air and water), and discrimination and violence. Climate justice researchers largely explore the relationships among nation-states within the context of climate science, and the responsibilities and international agreements that may be brought to bear to mitigate intensification of climate change across the globe. In particular, discussion tends to be focused on the distribution of climate change consequences and responsibilities between what are often described as developed and developing nations (Wallimann-Helmer, 2015).

Aligned with Pulido's (2015) aim of illuminating how race and racism play out in all social and socio-spatial contexts, climate justice might explore how the structures and ideologies of class and race oppression obstruct action toward mitigating climate change. One outcome may be that reporting on developments such as the increase in frequency and magnitude of storms, vulnerability associated with rising sea levels, and heightened probability of drought with its regional and global impact on food and water takes on new meaning. Specifically, an understanding how deficit narratives about the geographic Other—particularly in regions labeled underdeveloped—are embedded within narratives of class and race, and serve to disassociate the beneficiaries of unequal resource exploitation from the immediate consequences of climate change.

Insights from Critical Ecopedagogy

Aligned with the aims of Environmental and Climate Justice scholars, Critical Ecopedagogy approaches explore the connection between structures of power and environmental destruction. In his book *Critical Pedagogy, Ecoliteracy & Planetary Crisis,* Richard Kahn (2010) strives to find common ground between *environmental education* and the more politically engaged *critical pedagogy*. Kahn's primary concern is a rational response to the ecological crisis currently impacting all regions of the world. As a researcher and activist working on environmental issues, Kahn examines the complexity of factors that contribute to the growing threat to ecological stability. For Kahn, there can be no solution outside the domains of politics and discourse, and the eradication of species and destruction of ecosystems are closely related to the domination and exploitation of people (p. 6). Kahn's critique of the underlying conditions that give rise to the global ecological crisis begins with a globalized economics of exploitation (neo-liberalism) and extends to the political, educational, and media networks offering ideological cover for the damaging consequences of the economic program (p. 3).

Following the lead of Edgar Gonzalez-Gaudiano, an influential ecopedagogy advocate, and inspired by the work of Paulo Freire, Kahn (2010) envisions a pedagogical program that "seeks knowledge of how the environmental factors that contribute to disease, famine, unemployment, crime, social conflict, political repression, and other forms of sexual, ethnic, or religious violence can be examined as complex social and economic problems deserving everyone's attention" (p. 14). Kahn recommends a problem-posing approach to education, in which learners trace patterns of social, political, economic, and ecological factors that shape the challenges identified in local and global contexts (p. 14).

Kahn (2010) draws heavily on the works of Herbert Marcuse and Ivan Illich to develop a theoretical grounding for Ecopedagogy. Kahn argues for the need to challenge and reconstruct "an anthropocentric world view," that he characterized as "a matrix of domination;" a "global technocapitalist infrastructure;" an "unsustainable reductionist, and antidemocratic model of institutional science." He also argues for confronting the "marginalization and repression of pro-ecological resistance" (p. 9). For Kahn, the sum of these interventions begins to describe the basis of ecoliteracy that ought to drive environmental education: an ecoliteracy "requiring critical knowledge of the dialectical relationship between mainstream lifestyle and the dominant social structure. ... [A] much more radical and complex form of ecoliteracy than is possessed by the population at large" (p. 6). Kahn (2010) articulates a particular perspective toward understanding oppression, one that focuses on patterns of institutional, structural, and interpersonal oppression together with the commodification and consequence of Land use and ecological destruction.

Insights from Land Education

As discussed above, the question of Land is key to addressing both social and ecological justice theory. The research areas of critical theory, critical race theory, decolonization, and environmental justice are interwoven with the layered realities and meanings of Land: how it is understood, accessed, inhabited, used, altered, controlled, and represented. This emphasis on Land in social and ecological justice work is reflected in a growing body of scholarship published under the heading Land education. Land education takes up place-based education's goal of interrogating "What is happening here? What happened here? What should happen here? What needs to be transformed, restored, or conserved in this place?" (Greenwood, 2009, p. 279). It does so, however, through the lens of decolonization, which has a clear connection to analyses of race and class.

In addition to the myriad ways that unequal power manifests around Land as racism, classism, gender discrimination, surveillance, resource extraction, gentrification, industrial waste zoning, etc. (Bullard, 1990; Leonardo, 2009; Pulido, 2015), oppression within colonial spaces also involves the interplay of distinct ontological and cosmological orientations to Land. Kathi Wilson and Evelyn Peters (2005) reveal the paradoxical nature of place for Indigenous Peoples subjected to colonization—as both a source of strength and a space of alienation, restriction, and struggle. The authors emphasize that *place* does not come into being outside of the forces of conflict and violence—just as identity and sense of belonging do not exist outside of these influences (Wilson & Peters, 2005).

Eve Tuck, Marcia McKenzie, and Kate McCoy (2014) see similar tensions of relationality to place within settler colonial contexts as having bearing upon educational design in general and place-based pedagogy in particular. The authors illuminate the distinguishing and operational features of Land education in the introduction to a special issue of *Environmental Education Research.* As a nascent field, Land education is distinct from place-based education in terms of how place (Land) is understood in relation to political-historical and cosmological framings. Centering on global regions established through mechanisms of colonialism, the territories within which education takes place must be read through this history that is so dominated by the project of colonialism. In their discussion of settler colonialism, Tuck et al. (2014) discuss the ways that settler inhabitants resist understanding and acknowledging that within the knowledge systems of Indigenous Peoples is a "pre-existing ontological and cosmological relationship" with Land (p. 7). This gap in and aversion to understanding on the part of settlers creates boundaries of identity and presence within place, and serves as a foundation of the ongoing settler colonial project. The impetus to open educational design to Indigenous conceptions of Land also involves the enrichment of meaning of Land, as the idea of " 'Land' refers not just to its materiality, but also its 'spiritual, emotional, and intellectual aspects' " (Styres, Haig-Brown, and Blimkie, cited in Tuck, McKenzie, & McCoy, 2014, p. 9). This multidimensional approach to exploring what is means to be present on and have a relationship with Land has important implications for social and ecological justice learning and action.

Along with the assumption of a territorial worldview through which place is imagined, Tuck, McKenzie, and McCoy (2014) identify the settler colonial presumption of a singular "futurity" as an extension of settler nation-building consciousness (p. 11). Tuck at al. discuss themes of Indigenous agency and resistance in the context of a recent emphasis on place and place-based educa-tion as a response to global social and ecological crises (p. 15). Land educa-tion both draws on Indigenous knowledge as a key resource for responding to such crises, and energizes the possibility of a plural futurity that is more amenable to just and prosperous arrangements for both Indigenous and non-Indigenous inhabitants.

Discussion: Identifying Patterns in the Theory

Given the interdisciplinary nature of this approach to reading relevant liter-atures, it seems helpful to develop the analysis around a holistic framework. Fritjof Capra's (2005) Theory of Living Systems provides a useful model for engaging an integrated discussion of this diverse and dynamic body of work. In Capra's Theory of Living Systems, the first task is to identify a network of interconnected elements that make up the system being researched. The elements comprising the network are understood to act upon one another in what some theorists term dynamic balance (p. 28). In an (eco)system, the network might be established, for instance, through looking at soil, plants, animals, insects, water, climate, air, human activities, etc. In order to under-stand the system, it is necessary to examine each of these interconnected components both individually and in a relationship with the others. As this discussion has unfolded, it has become increasingly clear that the path to understanding what is meant by social and ecological justice—or simply what we wish to see improved in the world—follows a similar model. The path to understanding, therefore, involves gaining knowledge of multiple foun-dations of social and ecological justice, seeking patterns found across the diverse research areas dedicated to guiding learning and action. When a spe-cific problem is established as a starting point for analysis, racism for instance, efforts toward effecting change might be perceived to diverge or even com-pete with efforts of those who set colonization, heterosexism, or environmen-tal sustainability as the core problem. Referencing a network model, it might be anticipated that a research program centrally concerned with racism will interpret research around colonialism, heterosexism, and environmental sus-tainability as potentially enhancing understandings and interventions related to racism as these phenomena are intricately interwoven.

Although the above review of social and ecological justice theory is merely a sampling of the extensive work that has been done, it allows us to look for useful overarching or shared and recurring approaches to com-plex problems, and to contemplate how such framings might inform learning and action. One core pattern that can be clearly identified is the necessity to dig deeply into the construction, reproduction, and inculcation of dominant group identities (Butler, 1990; Calderon, 2014; Fanon, 1963; Kumashiro, 2009; Leonardo, 2009; Pulido, 2015). As Kumashiro (2009) has argued, "The reason we fail to do more to challenge oppression is not merely that we do not know enough about oppression, but also that we often do not *want* to know more about oppression. It is not our lack of knowledge but our resistance to knowledge and our desire for ignorance that often prevent us from changing the oppressive status quo" (p. 27). Often this resistance to knowledge is anchored in a deep yet largely unexamined investment in narratives based in dominant positionalities. An integrated understanding of social and ecological justice will thus require ongoing, systematic examination of how investments in dominant narratives play out through our performativity (Butler, 1990); and how such investments by others play out through performativity directed towards us.

In addition, through more deeply examining the construction and reproduction of dominant-group narratives, commonalities may be discerned with the ways that identity constructs such as talent, success, and social value (all of which are tied to class) are performed through codes of consumption. Just as our identities are bound up with narratives of social group membership, our identities are also heavily mediated by our ability to purchase "goods" that will serve to telegraph our status in a competitive society. Due to this integral relationship between consumption, status, and identity, it becomes difficult to internalize new knowledge about how our consumption contributes to the present ecological crisis, and that mitigating the crisis will require departing from and subverting the dominant codes of consumer culture (Kahn, 2010; Kumashiro, 2000).

A second pattern or repeated focus that has emerged in this review is the centrality of Land as a locus for understanding and working towards social and ecological justice. Throughout the world, the desire for Land and the wealth it provides have been a motivation for violence, as well as the ideological and moral fabrications required to maintain and disassociate from such horrific acts (Fanon, 1963; LaRocque, 2010). While the origins of all oppression cannot be directly linked to the appropriation of Land, all forms of oppression coalesce within dominant cultural identities that are situated on and rooted in the Land (Calderon, 2014; Tuck et al., 2014). An integrated theory of social and ecological justice centring on Land therefore asks the questions: through what processes of annexation, violence, exploitation, fabrication, and consumption have we come to this social and ecological reality in this time and place? And, in what ways do our actions in this time and place maintain these processes both locally and globally? In order to respond to these questions, we can see the value (and necessity) of drawing on insights from critical theory, critical race theory, feminist theory, queer theory, decolonization, environmental justice, climate justice, and Land education.

Through engaging these critical questions about our presence on the Land, we can also recognize the importance of digging deeply into the framing and reproduction of dominant narratives. For many researchers, engaging such questions means seeing our identities as entangled with the problems we wish to interrogate, and our very presence on the Land as rooted in a history and futurity defined by social and ecological injustice. By centring Land, the objective shifts from identifying a specific problem and seeking solutions within a specific research realm, to observing, naming, and challenging the ways colonization, racism, classism, heteropatriarchy, anthropocentrism, and all other forms of discrimination and violence are embedded within and play out in our relationships with one another and with the Land. Conscious, deliberate engagement with such critical questions then becomes a formative part of who we are and how we live together.

This type of analysis that is advanced by many of the subfields and areas of inquiry reviewed here is sometimes labeled intersectionality, and is increasingly a requisite for social and ecological justice theorizing (Sensoy & DiAngelo, 2012). Such analysis may also require us to see ourselves paradoxically (Kumashiro, 2000, 2009). Kumashiro (2000) encourages justice researchers to be open to the possibility of being influenced by and reproducing oppression, while simultaneously striving to transform it. Whether or not we can relate to a conceptualization of paradoxical identities embedded in systems of unequal power and oppression, we must include in our methodological approach a deliberate examination of potential "blind spots", and place value on identifying new forms of analysis. Understanding "the problem" in social and ecological justice is extremely complicated, and our interpretation of the problem will always be partial. This is in part due to our identities being bound up in narratives and structures of oppression, which necessitates a constant process of critical reflection (Kumashiro, 2009). What we might take away from this acknowledgment of the messiness and complexity of social and ecological justice work is a commitment to posing questions that challenge our own beliefs, values, and identities.

Reflecting on what one might learn from these literatures about how to pursue social and ecological justice in a changing and complex world, it becomes apparent that generating a useful understanding of what is going wrong requires an interdisciplinary framework that goes beyond aggregating an array of discrete methodologies. It requires identifying patterns and symmetries within the diverse research areas, empowering analyses able to articulate and confront the multiple factors that give rise to our problems. Wendell Berry (2005) states that "a bad solution is bad, then, because it acts destructively upon the larger patterns in which it is contained. ... A good solution is good because it is in harmony with those larger patterns" (p. 33). When we ask what is going wrong and what we might do to make a difference, it seems promising to approach social and ecological justice through a practice of solving for pattern. We need to become more able to address multiple, interlocking problems simultaneously, and to better recognize the contributions of those around us who are working toward similar visions of social and ecological justice.

Note

1. I use the term ecological rather than environmental throughout this chapter to emphasize the interrelationships of systems that establish the foundations for life, as well as the mutually-destructive consequences of those interrelationships being corrupted.

References

Berry, W. (2005). Solving for pattern. In M. K. Stone, Z. Barlow, & F. Capra (Eds.), *Ecological literacy: Education our children for a sustainable world* (pp. 30–40). San Francisco, CA: Sierra Club Books.

Bourdieu, P. (1984). *Distinction: A social critique of the judgment of taste.* New York: Routledge.

Bryant, B., & Mohai, P. (Eds.). (1992). *Race and the incidence of environmental hazards.* Boulder, CO: Westview Press.

Bullard, R. (1990). *Dumping in dixie: Race, class and environmental quality.* Boulder, CO: Westview Press.

Butler, J. (1990). *Gender trouble: Feminism and the subversion of identity.* New York: Routledge.

Calderon, D. (2014). Speaking back to manifest destinies: A land education-based approach to critical curriculum inquiry. *Environmental Education Research, 20*(1), 24–36.

Capra, F. (2005). Speaking nature's language: Principles for sustainability. In M. K. Stone, Z. Barlow, & F. Capra (Eds.), *Ecological literacy: Education our children for a sustainable world* (pp. 18–29). San Francisco, CA: Sierra Club Books.

Dimick, A. (2012). Student empowerment in the environmental science classroom: Toward a framework for social justice science education. *Science Education, 96*(9), 990–1012.

Fanon, F. (1963). *The wretched of the earth.* New York: Grove Press.

Giroux, H. (2009). Critical theory and educational practice. In A. Darder, M. P. Baltodano, & R. D. Torres (Eds.), *The critical pedagogy reader* (pp. 27–51). New York: Routledge.

Gosine, A., & Teelucksingh, C. (2008). *Environmental justice and racism in Canada: An introduction.* Toronto, ON: Emond Montgomery Publications Limited.

Greenwood, D. (2003). The best of both worlds: A critical pedagogy of place. *Educational Researcher, 32*(4), 3–12.

Greenwood, D. (2009). Place: The nexus of geography and culture. In M. McKenzie, P. Hart, H., Bai, & B. Jickling (Eds.), *Fields of green: Restorying culture, environment, and education* (pp. 271–282). Cresskill, NJ: Hampton Press Inc.

hooks, b. (1994). *Teaching to transgress: Education as the practice of freedom.* New York: Routledge.

Kahn, R. (2010). *Critical Pedagogy, ecoliteracy, & planetary crisis: The ecopedagogy movement.* New York: Peter Lang Publishing Inc.

Kirsch, M. K. (2000). *Queer theory and social change.* New York: Routledge.

Knutti, R., & Rogelj, J. (2015). The legacy of our CO_2 emissions: A clash of scientific facts, policies and ethics. *Climate Change, 133,* 361–373.

Kumashiro, K. (2000). Toward a theory of anti-oppressive education. *Review of Educational Research, 70*(1), 25–53.

Kumashiro, K. (2009). *Against common Sense: Teaching and learning toward social justice.* New York: Routledge.

Ladson-Billings, G. (2009). Just what is critical race theory and what's it doing in a *nice* field like education?. In E. Taylor, D. Gillborn, & G. Ladson-Billings (Eds.), *Foundations of critical race theory in education* (pp. 17–26). New York: Routledge.

Ladson-Billings, G., & Tate, W. (2006). Toward a critical race theory of education. In A. D. Dixson, & C. K. Rousseau (Eds.), *Critical race theory in education: All God's children got a song* (pp. 11–30). New York: Routledge.

LaRocque, E. (2010). *When the other is me: Native resistance discourse.* Winnipeg, MB: University of Manitoba Press.

Leonardo, Z. (2009). *Race, whiteness, and education.* New York: Routledge.

McLaren, P., & Kincheloe, J. (Eds.). (2007) *Critical Pedagogy: Where are we now?* New York: Peter Lang.

Memmi, A. (2000). *Racism.* Minneapolis, MN: University of Minnesota Press.

Nightingale, A. J., Eriksen, S., Taylor, M., Forsyth, T., Pelling, M., Newsham, A., Boyd, E., Brown, K., Harvey, B., Jones, L., Kerr, R. B., Mehta, L., Naess, L. O., Ockwell, D., Scoones, I., Tanner, T., & Whitfield, S. (2020). Beyond technical fixes: Climate Solutions and the Great Derangement. *Climate and Development, 12*(4), 343–352.

Nolet, V. (2016). *Education for sustainability: Principles and practices for teachers.* New York: Routledge, Taylor & Francis Group.

Palmer, J. (1998). *Environmental education in the 21st century: Theory, practice, progress and promise.* New York: Routledge.

Peet, R. (2007). *Geography of power: The making of global economic policy.* New York: Zed Books.

Pulido, L. (2000). Rethinking environmental racism: White privilege and urban development in southern California. *Annals of the Association of American Geographers, 90*(1), 12–40.

Pulido, L. (2015). Geographies of race and ethnicity I: White supremacy vs white privilege in environmental racism research. *Progress in Human Geography, 39*(6), 809–817.

Rose, J., & Cachelin, A. (2014). Critical sustainability in outdoor education: Connection to place as a means to social justice and ecological integrity. *Taproot, 23*(1), 7–16.

Roser, D., Huggel, C., Ohndorf, M., & Wallimann-Helmer, I. (2015). Advancing the interdisciplinary dialogue on climate justice. *Climate Change, 133,* 349–359.

Sensoy, O., & DiAngelo, R. (2012). *Is everyone really equal? An introduction to key concepts in social justice education.* New York: Teachers College Press.

Sium, A., Desai, C., & Ritskes, E. (2012). Towards the 'tangible unknown': Decolonization and the Indigenous future. *Decolonization: Indigeneity, Education & Society, 1*(1), I–XIII.

Tuck, E., McKenzie, M., & McCoy, K. (2014). Land education: Indigenous, post-colonial, and decolonizing perspectives on place and environmental education research. *Environmental Education Research, 20*(1), 1–23.UCC *see* United Church of Christ's Commission for Racial Justice.

United Church of Christ's Commission for Racial Justice. (1987). *Toxic waste and race in the United State: A national report on the racial and socio-economic characteristics of communities surrounding hazardous wastes sites.* New York: United Church of Christ.

United Nations. (1987). *Our common future: The world commission on environment and development.* Oxford University Press.

United Nations. (2015). *Sustainable development goals: 17 goals to transform the world.* Retrieved from <http://www.un.org/sustainabledevelopment/sustainable-development-goals/>.

Wallimann-Helmer, I. (2015). Justice for climate loss and damage. *Climate Change, 133,* 469–480.

Warren, K., Roberts, N. S., Breunig, M., & Alvarez, M. A. (2014). Social justice in outdoor experiential education: A state of knowledge review. *Journal of Experiential Education, 37*(1), 89–103.

Wilson, K., & Peters, E. (2005) "You can make a place for it": Remapping urban First Nations spaces of identity, *Society and Space, 23,* 395–413.

"Water Deep" (2020)
Sky McKenzie

2. Mutually Entangled Futures in/for Environmental Education

Kathryn Riley

Living and working on the traditional territories of the Neyinowak Inniwak (Cree) and Métis peoples.

Opening to a (Re)Storying of Environmental Education

Society is bombarded by narratives of ecological, social, and cultural exploitation, domination, and objectification in these times of the Anthropocene.[1] As environmental education (EE) is enmeshed within a Western trajectory of cultural norms and practices (Tuck & McKenzie, 2015), it is consequently directed by anthropocentric (dominance of the human species) and humancentric (dominance of the human self) logics. These logics manifest in globalizing discourses of 'common sense' that diffuse and obscure local histories, practices, and policies; neoliberal ideologies that purport independent and individual autonomy to enact behavior change for more sustainable societies and, thus, positions the individual as separate and discrete from societies and ecologies in focus; modes of appropriation set within extractive capitalism commercializing all of planet Earth; and ongoing colonial relations to Land[2] as characterized by conquest and genocide.

Situated in the posthumanist turn of educational research, the purpose of this chapter is to trouble the Western trajectory of cultural norms and energize and materialize the furthering of ethical pathways away from anthropocentric and humancentric logics within dominant EE research (Adsit-Morris, 2017; Clarke & Mcphie, 2020; Gough & Gough, 2016; Malone et al., 2020). The posthumanist turn is derived from the poststructuralist school of thought that seeks to challenge binary classifications subjugating and marginalizing those deemed as 'Other'[3]. However, as poststructuralism meets anthropocentric and humancentric limits through its focus on the culturally oriented, discursive (social) human (Barad, 2007), posthumanist perspectives move beyond a purely discursive gaze focused upon social dynamics to also take up affective materiality. Affective materiality is the nonconscious and pre-personal intensities of bodily states, in which senses-sensing[4] generate feelings and modes of thought and action *through* relational intraactions with 'Other(s)' (Braidotti, 2013; Massumi, 2015; Rautio, 2017). This (re)orients agency to be understood as distinct in *relation to*, rather than existing as an individual element, as something humans 'have' (as the term interaction would propose) (Barad, 2007). As such, affective materiality provides the conditions of possibility to embody the inextricable interdependence and interconnection between human/Earth relationships in EE.

Thinking/doing-with/through posthumanist perspectives in this chapter, I provide a non-dualistic mapping of difference between myself, as a White/Western/Settler[5] to Canada and a grey wolf (*Canis lupus*)[6] named Buddy (Braidotti, 2011; Lenz Taguchi & Palmer, 2013). Through the intensities of affective materiality generated from this Settler/Wolf story, I illustrate a (re)configuring of oppositional and dualistic difference and a (re)drawing of subject/object boundaries, while also challenging the subsequent hierarchical ordering based on this difference (Fine, 1998). I was (and am) still very attached to my 'humanness', as this is not something I cannot, nor something I want to, escape from or transcend; particularly since posthumanist approaches to worldmaking are enacted from the grounded, lived, embodied, and embedded (micro) politics of location. However, as I will explore in this chapter, in wandering, wondering, gathering, attuning, and attending to new seeds/stories through the intensities of affective materiality, I began to embody a new and different logic that decentered and expanded a sense of humanness.

Wandering, Wanderlust, and the Wayfaring Backpacker

I come from Country traditionally known as Boon wurrung Country, or what settler colonial Australia calls southern Victoria. In 2015, I (im)migrated to another settler colonial society in Canada, imbued with a colonial imaginary in which I storied place from settler inscriptions (Nxumalo & Cedillo, 2017). For example, glossy travel magazines and a plethora of social media sites that portrayed Canadian landscapes as vast, rugged, and beautiful prompted a romantic wanderlust. I romantically envisioned Canada as a sprawling wilderness full of thick, bottle green forests, and beige sandy beaches fringing aqua flows of meandering rivers and deep alpine lakes. I conjured romantic imaginaries of all the multispecies living within these pristine landscapes: trees of Jack pine (*Pinus banksiana*), white spruce (*Picea glauca*), blue spruce (*Picea pungens*), balsam fir (*Abies balsamea*), tamarack (*Larix laricina*), silver maples (*Acer saacharinum*), trembling aspen (*Populus tremuloides*), white birch (*Betula papyrifera*), balsam/black poplars (*Populus balsamifera*), and wolves (*Canis lupus*), bald eagles (*Haliaeetus leucocephalus*), beavers (*Castor canadensis*), grizzly bears (*Ursus arctos horribilis*), black bears (*Ursus americanus*), cougars (*Puma concolor*), white-tail deer (*Odocoileus virginianus*), elk (*Cervus canadensis*), moose (*Alces alces*): the hunters and the hunted.

While I understand that Canadian flora and fauna host many, many more species than what I have included here (including annoying pests), it is no mistake these 'charismatic mega-fauna' (as coined by E.O. Wilson in the late twentieth century with reference to media-popular animals) were at the forefront of my colonial imaginary. Steeped in naïve romanticism, this colonial imaginary was not at all set within the grounded sociocultural, political, and environmental realities of Canada as a settler and capitalist nation. This became more prominent as I 'settled' in the prairie province of Saskatchewan, one of thirteen provinces and territories in Canada.

Like so many other places in the world, Saskatchewan bears the effects of ecological, social, and cultural crises narratives, given that the province's ecological footprint is above the Canadian average (City of Saskatoon, 2014). Saskatchewan is also home to the largest potash mining corporation, *Nutrien*, formally known as the Potash Corporation of Saskatchewan before its amalgamation with Calgary-based Agrium in early 2018. Extensively mined and manufactured throughout the Canadian prairies, potash is a salt containing potassium. It acts as an agricultural fertilizer to improve water retention, yield, nutrient value, taste, color, texture, and disease resistance food crops (Western Potash Corp., 2018). According to the Western Potash Corp. (2018), as Canada exports 95% of its potash to over 50 countries around the world, it is literally 'feeding the world'. While *Nutrien* has an enterprise value of 36 billion US$ (CBC, 21 June 2017, n.p.), the widespread and intrusive practices of mining potash are not without environmental and sociocultural impacts. These impacts include changes to the landscape, water contamination, excessive water consumption, and air pollution, and of course, the abandonment of multivocal approaches to Land management that includes Indigenous sovereignty (UNEP, 2001). In this way, economic prosperity is often pitted against these long-term environmental impacts, with media rhetoric spurring public debate regarding pervasive ecological tolls.

Treaties Two (1871), Four (1874), Five (1875), Six (1876), Eight (1899), and Ten (1906) encompass the province of Saskatchewan (Government of Canada, 2010). These treaties are a formal agreement between the Indigenous people on that territory and the Crown, in which both parties must fulfill obligations and expectations through "mutually beneficial arrangements that guarantee a co-existence between the treaty parties" (office of the Treaty Commissioner, 2018a, n.p.). Seventy-two First Nations live within Saskatchewan's borders, making up five linguistic groups of Cree, Dakota, Dene (Chipewyan), Nakota (Assiniboine), and Saulteaux (office of the Treaty Commissioner, 2018b). After Winnipeg in the province of Manitoba, the cities of Regina and Saskatoon in Saskatchewan have the second and third largest populations of First Nations in Canada (City of Saskatoon, 2014). Treaty 6, which expands Saskatchewan and Alberta, is also the homeland of the Métis.

In my first couple of months living in Saskatchewan I was drawn to the 'spiritual' community and my new-found friends invited me to many different Indigenous rituals and events, including drum circles, sweat lodges, Powwows, and smudging ceremonies. Yet what struck me with visceral intensity as I enthusiastically attended these rituals and events, was the idea that they acted as some sort of White/Western/Settler fantasy escape, in which Indigenous ontologies of Land[7] and culture were romanticized through a colonial imaginary. Juxtaposed against Saskatchewan's political, economic, social, and environmental landscapes bearing the impacts of these times of the Anthropocene, it became starkly evident I was, indeed, contributing to globalizing discourses and neoliberal and capitalist spinning wheels of a commercialized, colonized, and exploited Land and peoples.

This was further illuminated through many different conversations I had with people, in which I was referred to as an expat, not an (im)migrant. I was not inculcated with racial differentials. Rather, through my White/Western/Settler privileges, I held the necessary currency to (im)migrate from Australia to Canada out of choice, inspired by the romantic notions of a colonial imaginary. As Peggy McIntosh (1988) wrote with reference to White privilege, "Power from unearned privilege can look like strength when it is in fact permission to escape or to dominate" (p. 34). Because I was an outsider with enough privilege to 'claim' Canada as my new home, not only was I enacting forces of globalization (mobilized through my capacity to board the Boeing 747 and fly across the Pacific Ocean), neoliberalism (mobilized through politics of identity as imbued with independent and individual autonomy), capitalism (mobilized through opportunities to purchase a discounted flight), and colonialism (mobilized through assumptions I could readily access the Land and peoples of Saskatchewan). Then, I had a chance encounter with a grey wolf, named Buddy, living at the local Saskatoon Zoo and Forestry Farm. I felt the tug of the colonial imaginary when I first met Buddy, imagining this 'symbol' of Canadian wildness living alongside, and with, the cultural practices of the First Peoples of Canada. Yet, as I experimented with post-humanist perspectives and the role of affective materiality to 'unsettle' my Settler self, I looked for different ways to relate with Land and peoples of Saskatchewan; ways that sought to trouble White/Western/Settler privileges as described above. I was not interested to replace colonial imaginaries but as affective materiality 'marked' the body with vibrancy and liveliness, colonial imaginaries were storied alongside the (micro) politics of location as a White/Western/Settler in these times of the Anthropocene.

A Settler/Wolf Story

Figure 2.1: Buddy at the Saskatoon Zoo and Forestry Farm

Watching Buddy (shown in Figure 2.1) walk the perimeter of his enclosure, he was the epitome of wild to me. I was in awe of Buddy as I studied his strong paws, gracefully yet purposefully padding the snow; his knowing ears tuned forwards, alert for danger. His eyes appeared wise, as his thick coat bristled and stiffened in the January wind. Western cultural narratives of the wolf are rooted in the Middle Ages of Europe, in which people superstitiously believed in werewolves (Wolfcountry, n.d.). Encouraging a fear of wolves, this has been played out in the literature of children's fairy tales and fables, namely Charles Perrault's seventeenth century publication of *Little Red Riding Hood*, and *The Three Little Pigs*, retold in 1922 by Flora Annie Steel. These stories of humans distancing themselves from wolves were very familiar to me, growing up with them as a child. Within Indigenous ontologies of Land and culture, however, the wolf, is positioned in relational reciprocity (Coulthard, 2014; Deloria, 1999; Martin, 2017; Watts, 2013). As an apex predator, wolves are seen as a role model through their demonstrations of finely-tuned and complex social systems enabling wise, powerful, and instinctive hunting (Northern Lights Wolf Centre, 2018).

Through ongoing and pervasive colonial imaginaries that dualistically position humans over nature, Indigenous ontologies of Land and culture are often romanticized as 'better' or more ecologically appropriate than Western worldviews (Morgensen, 2009; Nxumalo & Cedillo, 2017; Tuck & Yang, 2012; Wildcat, 2005). As Tracy Friedel (2011) wrote:

> Of the pernicious representations of Indigeneity today, none is more equivocal than the trope of 'the Ecological Indian'. Borne from nineteenth-century romantic primitivism, this White construction has become a prevalent signifier in the environmental realm, an ideal to which Canadians and others look today for a critique of Western institutions. (p. 534)

As White constructions of the 'Ecological Indian' are often adopted as a tool to challenge Western worldviews set in anthropocentric and human-centric logics, the modern-day Enlightenment project of colonialism is further exacerbated through the (re)appropriations of Indigenous ontologies of Land and culture. However, Indigenous ontologies of Land and culture are derived from entirely different epistemological and cosmological foundations to that of Western worldviews. This means they cannot be easily combined or absorbed into Western epistemology (Tuck et al., 2014). For example, human/Earth relational reciprocity within universalizing Western worldviews is typically set in romantic notions of nature, in which nature is deemed as a source of emotional identification (e.g., Rousseau-ean logics). In contrast, human/Earth relational reciprocity within Indigenous ontologies of Land and culture is typically a worldly given and, thus, set in realism (Wildcat, 2005). This is portrayed in the excerpt from Pat O'Riley and Peter Cole's (2009) poem below:

curving shapes of canoe and paddle dance
with the reflection of snow-capped mountains
on stillness of water

coyote and raven are exhausted from a long day's paddle
and head their canoe towards the shore
to rest for a bit on the cool moss
raven fluffs his feathers fluttering sighs
a cedar branch wavers in the last light

after taking a looong drink of water from the mountain stream
coyote rests against the roots of a thousand year old fir
you know raven we've been part of the land and sky
and other scapes so long and so intimately
that we don't often think about our relationship to them-it-those ones
since them-it-those ones was/is us
and us was/is them-it-those ones
without there ever being a twain to meet or not

raven yawns   yeah whatever. (p. 125).

Living out settler emplacement stories of a colonial imaginary through my encounter with Buddy, I inevitably romanticized, idealized, essentialized, and universalized Indigenous ontologies of Land and culture from White/Western/Settler privileges. As such, White/Western/Settler privileges worked to appropriate, and thus, supersede and silence Indigenous ontologies of Land and culture. This reinstates binary classifications through constructing 'Other(s)' in dichotomous ways (Tuck & Yang, 2014) and, thus, (re)affirms the need for new and different stories of human/Earth relationships in these times of the Anthropocene; stories that depart from anthropocentric and humancentric logics reifying supremacy of humans over nature, to stories that understand that humans, plants, animals, technological objects, energies, and histories all participate in a co-created and co-production of sociopolitical collectives (worldly realities) (Nxumalo and Cedillo (2017). These new and different stories are not imbued with discursive power differentials, in which some narratives have a more meaningful sense of validity within specific discursive contexts. Rather, as they are rooted in the co-constitutive materiality of the human body *with* nonhuman natures, they dismantle the idea of a separate and discrete human to suggest humans-in-relation as they emerge as part of the world (Barad & Gandorfer, 2021).

Thinking-with my encounter with Buddy in this way meant both Buddy and I were implicated, in one way or another, by what it meant to be alive in these times of the Anthropocene. I witness his curiosity, his inquisitiveness, as he looked out from his enclosure, like I witnessed my own curiosity and inquis- itiveness for what lie beyond the bounds of the zoo. For example, I noticed towering trees in the distance and wondered what might live within these trees. I witnessed Buddy's exhalations of breath in the freezing temperatures, like I witnessed my own. We both needed to inhale and exhale oxygen to survive and we both had cardiovascular and cardiorespiratory systems to enable this. And I witnessed Buddy wandering and pacing his enclosure as he seemingly looked for his alpha mate, Zeppelin, like I witnessed my own hand searching for my partner's arm in trying to keep warm and connected with my 'mate'. In these moments of senses-senses, lively, vibrant affective materiality caused bodily states that generated feelings of wonder, awe, joy, but also sadness, despair, guilt, and shame as to what it meant to be a White/ Western/Settler complicit within the colonization of Buddy's 'wildness'. Through affective materiality generating feelings and modes of thought and action, I was called to consider how my Settler practices not only discursively reinstated social hierarchies (between Settler and Indigenous peoples) but also how settler practices continue to disturb the living materiality of grey wolves (Kuhl, 2018).

As a keystone species, wolves play a crucial role in balancing the eco- system. However, there is ongoing debate as to how conservation strategies should manage wolves in Western Canada to address their dwindling pop- ulation. Given that wolves in Western Canada are a vulnerable species and are conceived as something to be 'environmentally managed', sociopolitical regimes pit wolf conservation strategies against practices of logging, hunting, and urban development. In this way, binary classifications of environmen- talism/anti-environmentalism are maintained and the environment is rein- stated as something 'out there', as a location external, distant, and outside of humans (Cronon, 1995).

Intentions underpinning Buddy's relocation from the wild to a human- managed, enclosed space at the Saskatoon Zoo and Forrestry Farm might be steeped in care-ethics, especially given Buddy was orphaned from his mother and pack as a pup. However, questions arise as to how Buddy's relocation to the zoo might influence people's experiences of captive animals in zoos. That is, how Buddy is discursively labeled as 'wild' through his animal-natures, and thus, othered in dualistic and oppositional difference to socioculturally oriented human-cultures (Fawcett, 2002). Through affective materiality, however, difference is no longer conceived of as dualistic and oppositional but affirmative. For example, through intraactions with Buddy, my White/Western/Settler self is (re)configured through co-created and co-constituted existences *with* Buddy. By no means does this resolve me of White/Western/Settler response-ability to globalizing discourses and neoliberal, capitalist, and colonial structures and systems in these times of the Anthropocene. To the contrary, as I cannot partition myself away from impoverished structures and systems through independent and individual autonomy, I am response-able to how I enact 'good relations' with 'Other(s)' (Liboiron, 2021; Taylor et al., 2013). That is, I am not response-able to relationships because relationships exist but *because* worldings are made up of dynamic and complex relational entanglements.

(Re)Configuring binary classifications through posthumanist perspectives does not involve romanticizing, idealizing, essentializing, or universalizing the Indigenous 'Other(s)' through appropriating or co-opting Indigenous ontologies of Land and culture rooted in reciprocal relationality. Rather, attention turns to White/Western/Settler-subjectivities as complicit in colonial realities of the settler state. As Billy-Ray Belcourt (2015) argued, "modern human-animal interactions are only possible because of and through the historic and ongoing erasure of Indigenous bodies and the emptying of Indigenous lands for settler-colonial expansion" (p. 3). This is understanding that any robust ecological discourse and examination of Land-based relations first needs to acknowledge how these relations have been colonized by settler practices in the erasure of differently located Indigenous knowledges as set within reciprocal relationships (Hern & Johal, 2018). In the final part of this chapter, I explore how my intraactions with Buddy generated different ways to live and learn between the borders of a colonial imaginary and the grounded realities of these times of the Anthropocene and between the borders of 'unsettling' a settler-self and longing to belong in Saskatchewan, and what this might mean for the field of EE.

Implications for Environmental Education

Affective materiality as generated from the grounded, lived, embodied, and emplaced (micro) politics of location offers new and different environmental pathways; pathways that understand we are all mutually entangled in co-created and co-constituted shared futures (Alaimo & Hekman, 2008); Taylor et al., 2013).

Through mutual entanglements, the discrete and separate human subject that views reality as external, objective, foundational, and stable (as understood through Cartesian Representational Knowing), breaks down to reveal differences and distinctions within a complex rhizomic assemblage of relations. Ontological differences between knowledge making generated in Cartesian Representational Knowing and worldmaking generated through rhizomic assessblages of relations is shown in Figures 2.2 and 2.3 below:

Figure 2.2: Cartesian Representational Knowing (Barad, 2007)

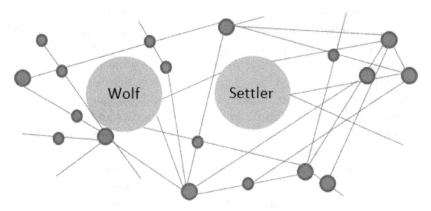

Figure 2.3: A Rhizomic Assemblage of Relations

Cartesian Representational Knowing exacerbates 'grand narratives' and 'master stories' that are common place in these times of the Anthropocene because it fails to recognize how grand narratives and master stories work to mediate truth claims and the pre-determined conceptions of knowledge Thus, Cartesian Representational Knowing is irreducible to sociocultural systems of dominant discourse (Rosiek & Gleason, 2017). Conversely, as demonstrated through the decentralized mapping of interconnected nodes (circles) in Figure 2.3, rhizomic assemblages of relations dismantle hierarchical positions of knowledge to reveal mutual (re)configurings *with* the world (Barad, 2007). In this way, rhizomic assemblages of relations take up the material (matter) as continuous with the discursive (social) rather than understanding matter as opposed to culture through matter/social binary classifications. As the unit of analysis turns to discursive *and* material forces stirring bodies into feelings and modes of thought and action, we open to a dense network of kin relations that are inextricably enmeshed and interconnected (Bennett, 2010).

The idea that individual subjectivities transform with/through relational intraactions within any given assemblage has significant implications for the field of EE, because it begets an anticolonial praxis that is rooted in the 'undoing' of White/Western/Settler subjectivities (Rautio, 2017). For example, through intraactions with Buddy, while I am still discursively positioned as a White/Western/Settler, as my body is pulled to feelings and modes of thought and action through senses-sensing with Buddy, I am also becoming unfamiliar with White/Western/Settler colonial imaginaries in becoming-wolf. This presents a paradox in posthumanist perspectives, because to enact change within impoverished structures and systems in these times of the Anthropocene, I first need to start from the (micro) politics of location as a White/Western/Settler to open to the possibilities to embody different modes of knowing/being/thinking/doing/feeling that is generated through, and because of, intraactions with 'Other(s)'. In this way, a sense of belonging is cultivated *because* of an 'unsettling' that is activated through dynamic, continual, and reiterative 'becoming-withs'. This is not diluting a sense of self; nor is it taking up 'Other(s)' as a utility or commodity through capitalist, neoliberal, and colonial tendencies to 'thingify' 'Other(s)'. Rather, understanding that we are re-made time and time again through relational intraactions with 'Other(s)', the idea of nomadic subjectivities (Braidotti, 2011) opens EE to a myriad of possibilities for new assemblages, new worldings, and new stories in/for the field.

I can only ever speak from my own subjectivities, in that I am not in any kind of ethical position to account for anyone/anything else's story but my own. In other words, I cannot speak for how affective materiality might 'mark' other bodies as it is impossible (and unethical) to suggest that someone/something can speak for someone/something else. This is something that is typically taken up through anthropocentric and humancentric logics (e.g., humans speaking for nature or animals; White/Western/Settlers speaking for Indigenous peoples). Yet as my story is mutually imbricated with you, the reader, it is my hope that through affective materiality 'marking' the body through intraactions with these pages, that your body might also be pulled to feelings and modes of thought and action that acknowledge an interdependent and interconnected 'togetherness relationship'. It is my hope that you do not turn away from the realities of these times of the Anthropocene but enact an opening to the conditions of possibility for sowing new seeds/stories for future (re)seeding and (re)worlding in these messy, confusing, and often terrifying times of the Anthropocene.

Notes

1. The Anthropocene refers to the present-day geological epoch and global movement, superseding what was known as the Holocene. Through the extent of ecological modification at the hands of the Anthropos, the Anthropocene marks severe discontinuities, in that while the planet might continue with or without us, the ecosystem and critter entanglements that we have become accustomed to may not (Haraway, 2016). Arriving at a critical juncture, there are still opportunities to (re)story how we understand and relate with planet Earth. To generate stories of hope about the kinds of futures that are created and cultivated *with* 'Other(s)', I therefore refer to the Anthropocene as a boundary rather than an epoch and, thus, take up the term 'in these times of the Anthropocene'.
2. 'Land' is capitalized in this chapter to extend beyond 'land' understood as an empty container that only includes the materiality of earth, rocks, and waterways, etc. Denoting a material/discursive relationship, Land is highly contextualized, culturally positioned, and threaded within an interconnected and interdependent assemblage of relations.
3. 'Other(s)' is in quotation marks throughout this chapter, because through a philosophical orientation of posthumanist perspective, 'Other(s)' only exists in a relational capacity, rather than as something separate and discrete (Barad, 2017).
4. Troubliing the idea that knowledge is 'out there' to be acquired through an understanding *of* the world, I depart from 'knowledge acquisition', 'meaning-making', and 'sense-making', as these terms depict a cognitive emphasis in generating new knowledge. Rather, I use the term 'senses-sensing' to depict embodied and relational forms of knowledge production (Ahenakew, 2016).
5. Within politics of 'White', 'Western', and 'Settler', these terms will be capitalized to indicate reference to specific social groups, also acknowledging the unstable nature of these categories in that 'White', 'Western', and 'Settler' groups possess fluid borders and heterogenous members. I also recognize that 'Settler' is a loaded term, given the vast differences in resources and privileges that non-Indigenous people have in coming to call Canada home.
6. Used to describe both the 'genus' and 'species' that categorize plants and animals, I include scientific Latin plant and animal names the first time I introduce plants and animals. This is to highlight the ubiquity of scientific discourse in these times of the Anthropocene (Haraway, 2016).
7. It is crucial to note that relationships to Land among Indigenous peoples are diverse, specific, and ungeneralisable begetting the importance to attend to contextualized, emplaced, and situated cultural practices over time (Tuck et al., 2014).

References

Adsit-Morris, C. (2017). Restorying environmental education: *Figurations, fictions, and feral subjectivities*. Switzerland: Palgrave MacMillan International.
Ahenakew, C. (2016). Grafting Indigenous ways of knowing onto non-Indigenous ways of being: The (underestimated) challenges of a decolonial imagination. *International Review of Qualitative Research, 9*(3), 323–340.

Alaimo, S., & Hekman, S. (2008). Introduction: Emerging models of materiality in feminist theory. In S. Alaimo & S. Hekman (Eds.), *Material feminisms*. Bloomington and Indianapolis: Indiana University Press.

Barad, K. (2007). *Meeting the universe halfway: Quantum physics and the entanglement of matter and meaning*. Durham & London: Duke University Press.

Barad, K. (2017). Troubling time/s and ecologies of nothingness: Re-turning, re-membering, and facing the incalculable. *New Formations, 92*, 56–86. < https://doi. org/10.3898/NEWF:92.05.2017>.

Barad, K., & Gandorfer, D. (2021). Political desirings: Yearnings for mattering (,) differently. *Theory & Event, 24*(1), 14–66.

Belcourt, B. R. (2015). Animal bodies, colonial subjects: (Re)Locating animality in deco-lonial thought. *Societies, 5*, 1–11.

Bennett, J. (2010). *Vibrant matter: A political ecology of things*. Durham, NC: Duke University Press.

Braidotti, R. (2011). *Nomadic theory: The portable Rosi Braidotti*. New York & West Sussex: Colombia University.

Braidotti, R. (2013). *The posthuman*. Cambridge, UK: Polity.

City of Saskatoon. (2014). *An overview of the demographic, economic, social, and environmental issues and trends in Saskatoon 2013–2014*. Retrieved from: <https://www.saskatoon.ca/sites/default/files/documents/asset-financial-management/finance-supply/office-finance-branch/2014–issues-and-trends.pdf>.

Clarke, D. A. G., & Mcphie, J. (2020). New materialisms and environmental education: Editorial, *Environmental Education Research, 26*(9–10), 255–1265. <https://doi.org/10.1080/13504622.2020.1828290>.

Coulthard, G. S. (2014). *Red skin, white masks: Rejecting the colonial politics of recognition*. Minneapolis: Minnesota.

Cronon, W. (1995). The trouble with wilderness; or, getting back to the wrong nature. In W. Cronon (Ed.), *Uncommon ground: Rethinking the human place in nature*. New York: WW Norton & Co.

Deloria, V. (1999). *Spirit & reason: The vine Deloria, Jr., reader*. Wheat Ridge, CO: Fulcrum Publishing.

Fawcett, L. (2002). Children's wild animal stories: Questioning inter-species bonds. *Canadian Journal of Environmental Education, 7*(2), 125–139. Retrieved from: <https://cjee.lakeheadu.ca/article/view/260>.

Fine, M. (1998). Working the hyphens: Reinventing self and other in qualitative research. In N. K. Denzin & Y. S. Lincoln (Eds.), *The landscape of qualitative research: Theories and issues* (pp. 130–155). Thousand Oaks, CA: Sage.

Friedel, T. L. (2011). Looking for learning in all the wrong places: Urban Native youths' cultured response to Western-oriented place-based learning. *International Journal of Qualitative Studies in Education, 24*(5), 531–546. <https://doi.org/10.1080/09518398.2011.600266>.

Gough, A. & Gough, N. (2016). Beyond cyborg subjectivities: Becoming-posthumanist educational researchers. *Educational Philosophy and Theory, 49*(11), 1–13. <https://doi.org/10.1080/00131857.2016.1174099>.

Government of Canada. (2010). *Indigenous and northern affairs Canada: Treaty research report – Treaty Six (1876).* Retrieved from: <https://www.aadnc-aandc.gc.ca/eng/1100100028706/1100100028708>.

Haraway, D. (2016). *Staying with the trouble: Making kin in the Chthulucene.* Durham & London: Duke University Press.

Hern, M., & Johal, A. (2018). *Global warming and the sweetness of life: A tar sands tale.* Cambridge, MA: MIT Press.

Kuhl, G. J. (2018). Sharing a world with wolves: Perspectives of educators working in wolf- focussed education. *Environmental Education Research.* <https://doi.org/10.1080/13504622.2018.1523370>.

Lenz Taguchi, H., & Palmer, A. (2013). A more 'livable' school? A diffractive analysis of the performative enactments of girls' ill-/well-being with(in) school environments. *Gender and Education, 25*(6), 671–687. <https://doi.org/10.1080/09540253.2013.829909>.

Liboiron, M. (2021). *Pollution is colonialism.* Durham & London: Duke University Press.

Malone, K., Tesar, M., & Arndt, S. (2020). *Theorising posthuman childhood studies.* Singapore: Springer Nature. <https://doi.org/10.1007/978-981-15-8175-5>.

Martin, B. (2017). Methodology is content: Indigenous approaches to research and knowledge. *Educational Philosophy and Theory, 49*, 1392–1400. <https://doi.org/10.1080/00131857.2017.1298034>.

Massumi, B. (2015). *Politics of affect.* Cambridge, UK: Polity Press.

McIntosh, P. (1988). *White privilege: Unpacking the invisible knapsack.* Retrieved from: <https://files.eric.ed.gov/fulltext/ED355141.pdf?utm_campaign=Revue%20#page=43>.

Morgensen, S. L. (2009). Un-settling settler desires. In Unsettling Minnesota Collective (Eds.), *Reflections and resources for deconstructing colonial mentality.* Minneapolis, MN: self-published source book. Retrieved from: <http://unsettlingminnesota.files.wordpress.com/2009/11/um_sourcebook_jan10_revision.pdf>.

Northern Lights Wolf Centre. (2018). *About wolves.* Retrieved from: <http://www.northernlightswildlife.com/about-wolves.php>.

Nxumalo, F., & Cedillo, S. (2017). Decolonizing place in early childhood studies: Thinking with Indigenous onto-epistemologies and Black feminist geographies. *Global Studies of Childhood, 7*(2), 99–112. <https://doi.org/10.1177/2043610617703831>.

Office of the Treaty Commissioner. (2018a). *About the treaties.* Retrieved from: <http://www.otc.ca/pages/about_the_treaties.html>.

Office of the Treaty Commissioner. (2018b). *First Nations in Saskatchewan.* Retrieved from: 2 May 2018, <http://www.otc.ca/ckfinder/userfiles/files/fnl_1100100020617_eng.pdf>.

O'Riley, P., & Cole, P. (2009). Coyote and raven talk about the land/scapes. In M. McKenzie, P. Hart, H. Bai & B. Jickling (Eds.), *Fields of green: Restorying culture, environment and education*. NJ: Hampton Press.

Rautio, P. (2017). A super wild story: Shared human-pigeon lives and the questions they beg. *Qualitative Inquiry, 23*(9), 722–731. <https://doi.org/10.1177/1077800417725353>.

Rosiek, C. J., & Gleason, T. (2017). Philosophy in research on teacher education: An onto-ethical turn. In J. Clandinin & H. Jukka (Eds.), *The Sage handbook of research on teacher education*. London: Sage.

Taylor, A., Giugni, M., & Blaise, M. (2013). Haraway's 'bag lady story-telling': Relocating childhood and learning within a 'post-human landscape'. *Discourse, 34*(1), 48–62. <https://doi.org/10.1080/01596306.2012.698863>.

Tuck, E., McCoy, K., & McKenzie, M. (2014). Land education: Indigenous, post-colonial, and decolonizing perspectives on place and environmental education research. *Environmental Education Research, 20*(1), 1–23. <https://doi.org/10.1080/13504 622.2013.877708>.

Tuck, E., McCoy, K., & McKenzie, M. (2014). Land education: Indigenous, post-colonial, and decolonizing perspectives on place and environmental education research. *Environmental Education Research, 20*(1), 1–23. <https://doi.org/10.1080/13504 622.2013.877708>.

Tuck, E., & McKenzie, M. (2015). *Place in research: Theory, methodology and methods*. New York & London: Routledge.

Tuck, E., & Yang, K.W. (2012). Decolonization is not a metaphor. *Decolonisation: Indigeneity, Education & Society, 1*(1), 1–40. Retrieved from: <https://jps.library. utoronto.ca/index.php/des/article/view/18630>.

Tuck, E., & Yang, K. W. (2014). 'R-words: Refusing research. In D. Paris & M. T. Winn (Eds.), *Humanizing research: Decolonizing qualitative inquiry with youth and communities* (pp. 223–248). Thousand Oaks, CA: SAGE.

UNEP *see* United Nations Environment Programme.

United Nations Environment Programme. (2001). *Environmental aspects of potassium and potash mining*. Retrieved from: <https://wedocs.unep.org/bitstream/handle/ 20.500.11822/8071/-Environmental%20Aspects%20of%20Phosphate%20and%20 Potash%20Mining-20011385.pdf?sequence=2>.

Watts, V. (2013). Indigenous place-thought and agency amongst humans and nonhumans (First Woman and Sky Woman go on a European world tour!). *Decolonization: Indigeneity, Education & Society, 2*(1).

Western Potash Corp. (2018). *What is potash?* Retrieved from: <https://www.westernpot ash.com/>.

Wildcat, D. R. (2005). Indigenizing the future: Why we must think spatially in the twenty-first century. *American Studies, 36*(3–4), 417–440. Retrieved from: <https://www. jstor.org/stable/40643906>.

Wolfcountry. (n.d.). *Myths, legends, and stories*. Retrieved from : <http://www.wolfcoun try.net/stories/>.

"Ahora Adentro" (2016)
Araceli Leon Torrijos

3. Decolonial Pedagogy, Agroecology, and Environmental Education: Repositioning Science Education in Rural Teacher Education in Brazil

MARCELO GULES BORGES

Living and working on the traditional territories of the Guarani, Xokleng and Kaingang peoples

Introduction

Brazil's rural education movement advocates that education in rural areas take place from and for rural peoples (Barbosa, 2017; Brasil, 2002, 2010; Munarim, 2008). The dialogue between agroecology and environmental education (EE) has taken place in Brazil's rural education movement as a core to its educational project (Leite & Conceição, 2020; Paim, 2016; Petri & Fonseca, 2020). Among other places, this dialogue plays out in the pedagogy of the *Licenciatura em Educação do Campo* (LEdC), a rural teacher education program created in 2007 and currently active in public universities in the country. Previous researchers have highlighted how this rural teacher education program is decolonial (Farias & Faleiro, 2020; Oliveira, 2019; Petri & Fonseca, 2020).

The LEdC is a program focused on training teachers for rural schools, that developed out of social movement struggles for rural education (started in the 1990s, Munarim, 2008), in opposition to what was called pedagogical ruralism in Brazil. Pedagogical ruralism emerged in the 1930s and remains today as a colonial ideal that defends curricular practices that subordinate rural identities and knowledges to urban. It proposes that communities settle in rural areas and have as their primary function to serve the city. Thus, rural workers continue to be excluded from the mechanisms of insertion in society and culture, even if, in some cases, they are completely integrated into the capital production model (Arroyo, 1982; Neto, 2016). In the educational context, the consequences of pedagogical ruralism are what Arroyo (1982, p. 2) called "cultural colonialism of the city over the countryside" through curriculum, methods and school content.

One of the pedagogical endeavors of the LEdC is to place the environment at the center of discussions, building a social and ecological education engaged with rural territories. The Brazilian rural education movement makes the LEdC a radical approach to environmental education. The history of EE in Brazil has consolidated over the decades into the construction of critical EE (Carvalho & Frizzo, 2016; Layrargues & Lima, 2014; Loureiro & Layrargues, 2013; Thiemann et al., 2018), which is widespread both in academia and in the public education system (formal and teacher education). Parallel to rural education, critical EE emerged from social and ecological movements in the 1970s in Latin America and continued to develop in the 1980s/1990s through the institutionalization of EE (for instance, the Brazilian National Policy for Environmental Education launched in 1999). Historically, the primary foundation of educational inspiration for rural education and EE comes from critical pedagogy and popular education produced in the context of social movements (e.g., the Landless Workers Movement [MST] on rural education and the thoughts of Paulo Freire).

More recently, rural education and critical EE (Andreoli & Borges, 2020) have been aligned with the decolonial project (Andrade et al. 2019; Kassiadou, 2018; Petri & Fonseca, 2020; Sánchez et al., 2020; Silva, 2018; Zeferino et al., 2019), advocating for a pedagogy with/for the environment based on the struggle for social and ecological justice. By proposing new situations, methods and practices that creatively invert social and environmental priority in educational practices, this movement considers critical interculturality (Mignolo & Walsh, 2018; Walsh, 2010). Centering critical interculturality allows the deconstruction of a certain western-modern logic that denies places, ways and practices of knowing and producing knowledges. It is in the search to unlearn, relearn and construct other grammars and body experiences that political action reinvents ways of naming and acting in decoloniality. As Walsh (2017) reminds us, it is precisely in the global South that decolonial pedagogy has emerged from community practices, configuring itself as a set of "pedagogies of resistance and reexistence, signal[ing] affirmation, hope, and life in the midst of conditions of negation, violence, death, destruction, and despair" (p. 369). In line with the notion of decolonial pedagogy, the LEdC allows science and EE to emerge in practice as teacher education based on agroecology. The concept of diversity in agroecology ties culture and environment to transform the notion of human development, radicalizing the idea of listening to others.

This chapter explores how agroecology and EE dialogue with science education at LEdC in Brazil. For this, I present to readers an experience in the training of rural teacher candidates in the South of the country, during my work as a professor at the Federal University of Santa Catarina (UFSC). Departing from an ethnographic experience (Rockwell, 2009) that involves supervision and observation (analysis of participant observation, fieldnotes, and teaching report) during the construction of practicum in rural public schools, I describe agroecology as part of the decolonial project for curricula and rural teacher education. First, I present the organization of the UFSC's particular manifestation of LEdC, exploring how agroecology structures the curriculum and teaching practices. Taking agroecology and EE as a decolonial strategy, the practicum narrative that follows describes the construction of an approach centered on interculturality in science education. I conclude by highlighting learning from the LEdC and my writing as a decolonial practice.

The Licenciatura em Educação do Campo *(LEdC) at the* Federal University of Santa Catarina *(UFSC)*

Since 2007, the *Licenciatura em Educação do Campo* (LEdC) has been in place in Brazilian federal universities as a public policy for rural teacher education "for and by the countryside" (Barbosa, 2017, p. 119). The LEdC program certifies post-secondary students to work in rural schools (Middle and High school) across different areas of scientific knowledge (i.e., Natural Sciences, Mathematics, Arts, Social Sciences). Its curriculum is structured by agroecology and alternating pedagogy. The undergraduate degree is based on the Principles of Rural Education (Brasil, 2010). One of the main foundations of these principles is the maintenance of formal schools and quality education in the rural context. Beyond this, these principles emphasize that rural education must include rural people aiming to achieve a more emancipatory and transformative education in order to prepare educators, who work with school subjects to promote the articulation of knowledges between schools and communities.

The LEdC proposal starts from a critical perspective that highlights the contradictions of society in its relationship with environments. For example, what precedes each practicum are critical analyses of the surrounding cultural, social and environmental issues, the school routine, the political-pedagogical projects, the classroom dynamics, the teaching proposals of the supervising teachers, the pedagogical materials (such as textbooks), dialogue with schoolteachers, conversations with students in that class, etc. LEdC proposes that scientific knowledge must be built in dialogue with other aspects of knowledge to support social transformation. Furthermore, this approach to education affirms that access to scientific knowledge and access to land are human rights (Brasil, 2010; Munarim, 2008).

The Federal University of Santa Catarina (UFSC) LEdC program was created in 2009 (UFSC, 2009), from the National Program to Support Rural Teacher Education (Procampo, 2008, 2009) and the National Program for Rural Education (Pronacampo, 2012) proposed by the Ministry of Education of Brazil. Currently, the program is part of a set of more than 40 degrees (Medeiros et al., 2020) that take place in public universities and technical institutes in Brazil (Molina, 2015, 2017; Molina & Hage, 2015) and set the objective of

> training educators to work in basic education, specifically for the final grades of elementary school and high school in rural schools. They will be able to manage educational processes and develop pedagogical strategies aimed at forming critical thinking, autonomous and creative human beings capable of producing solutions to issues inherent to their reality, linked to the social quality of the development of rural areas [...]. (UFSC, 2009, p. 4)

The UFSC curriculum is structured with regard to mandatory, complementary, optional and extracurricular activities (UFSC, 2009). The subjects are, in addition, studied on an alternating basis, between university time (TU) and community time (TC). During TU, students take courses and perform activities on the university campuses, and during TC, students take courses and do pedagogical activities in their communities. In the final two years of the program, the practicum takes place in public schools at the elementary and high school levels. During their practicum students go to the classroom (preferably in schools in their home communities) to design and implement a community project that encompasses science and environmental education, based on the study of the territory carried out in the first two years of the program. Thus, the theme of the project and the organization of teaching constitute a relevant focus for the entire school community and have a direct impact on the life of the place.

Specifically, in the third year of the program, the practicum takes place in elementary school (grade 6–9) and in the fourth year in high school (grade 10–12). In addition to focusing on their area of future practice as teachers (science education), students aim to deepen their scientific knowledge in order to provide rural schools with access to systematized knowledge. The LEdC pedagogy proposals focus on organizing rural schools with alternative methodologies, expanding the theory and practice of school routines to go beyond the classroom walls and transform society.

The practicum in each year is organized into observation activities (approximately eight hours) and teaching (twenty-four hours) carried out individually or by intervention groups. The practicum consists of carrying out, with the help of a university advisor and school supervisor, the construction of a teaching plan and lesson plans to be taught in elementary and high school. All this material and its application is developed from the reports of studies of the territory carried out in groups of students during the first and second years of the program. These territorial studies are informed by participant observation and collaborative work with the classroom and school where the teacher candidates will work.

As Dos Santos Araújo and Porto (2019) remind us, the alternating pedagogy is structuring and is fundamental to the characteristics of the LEdC practicum:

> In the alternating pedagogy, the practicum is a time to improve scientific, theoretical and practical knowledge, which must be carried out in order to strengthen the dialogue between the learning in University-time (UT) and the activities in community-time (CT) and the organization of pedagogical work in rural schools and in non-school spaces. (p. 4)

It is noteworthy that at each completed annual stage of the course, reports are drawn up and seminars are proposed by teacher candidates with the entire school community, faculty, and university. The intention of these seminars is to socialize and share experiences, as well as analyze the strengths and weaknesses of teaching and the community project. All stages of the practicum happen at the same school, with the purpose of gaining a greater approximation and understanding of the school's reality. However, most schools do not have primary and secondary education in the same institution in rural Brazil, often making it impossible for this dynamic to happen.

When proposing projects in schools during the practicum, LEdC students often face challenges. Some institutions are not used to working with alternative pedagogies and methodologies that promote interdisciplinarity and include social participation. This creates concerns for teachers who organize learning into separate disciplines, which is currently common in Brazilian public education. The practicum is mandatory for students and is designed in a format that allows them to apply the project either individually or collectively with their classmates. This form of organization requires thinking and building the division of clear stages and communication with students in schools. The collective work (university/teacher candidates/schools/classrooms) requires the involvement of the different subjects of the educational process, not only to put into practice a proposal but also to launch a relentless search for how to actualize more environmental and place-engaged practices (Alves & Faleiro, 2019).

Agroecology as Part of a Decolonial Strategy

In the process of defining themes and contents in teaching projects, which occurs critically and are always based on territory, the teacher candidates identify a primary issue for LEdC. This process is focused on how to bring a significant contribution to the transformation of rural communities during the practicum. This highlights the role of agroecology to attend to environmental issues, due to their ability to articulate grassroots issues involving the countryside and the daily lives of families and communities (cultural, social, and environmental aspects) from different scales (local/global).

Like any other concept, category, or field that arises from different cultural contexts, agroecology has become a place of dispute over meanings. Taking Agroecology as practice, movement, and science (Norder et al., 2016; Wezel et al. 2011), I agree with Leff (2002, p. 42) that it is constituted as a set of systematized practice, based on techniques and traditional knowledge (from native peoples and peasants) "who incorporate ecological principles and cultural values into agricultural practices that, over time, were de-ecologized and de-culturalized by the capitalization and technification of agriculture". As Leff points out,

> Agroecological knowledge is a constellation of dispersed knowledges, techniques, and practices that respond to the ecological, economic, technical and cultural conditions of each geography and each population. These knowledges and these practices are not unified around a science: the historical conditions of its production are articulated at different levels of theoretical production and political action, which pave the way for the application of its methods and for the implementation of its proposals. Agroecological knowledges are forged at the interface between cosmovisions, theories and practices. (Leff, 2002, p. 36)

Agroecology is eminently in opposition and reaction to agricultural models that are based on modern rationality, which take the other (nature) as a place to be controlled and dominated. This logic is the ultimate colonial expression of what Quijano (2000) called the *coloniality of power and knowledge.* In colonial logic/modern rationality, what is considered practical knowledge (or in practice), such as that of more sustainable agriculture, is deemed to be inferior. It is worth explaining that this last model is the result of the understanding of science and technology as "gifts with which the Western imperial powers had presented their colonies, being a fundamental part of the discourse on the civilizing mission. This discourse is at the base of the justification of the colonial-capitalist relationship, which has always sought to legitimize itself through the colonial version of the grammar of modernity" (Meneses, 2014, p. 91). Therein lies the decolonizing role that agroecology plays: it becomes a place of possibility with an integrative and relational role within rural environmental knowledges and practices, generating subsistence, education and future prospects for rural communities. A territorially-based agroecology works against agro-industrial and global market colonization processes that destroy localized identities and knowledges. It takes up practical knowledge that involves skills and ancestral practices of plowing, planting, harvesting, transforming and eating food. These practices shape ways of being and knowing in the relationship with the land, as a life support and meaning of existence for life in the countryside.

During the LEdC program, the position that agroecology assumes in the organization of pedagogical practices is evident. This is either because it is positioned in the core of the four years of the program as an axis to the curriculum or because it is positioned as an organizing theme of the practices in the schools where the internships take place. Figure 3.1 below highlights how agroecology is assumed as one of the transversal axes of the course:

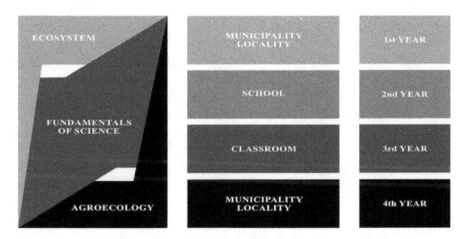

Figure 3.1: Axes of Curriculum (Ecosystem, Fundamentals of Science and Agroecology), Scales of Experience (Municipality/Locality, School, Classroom and Municipality/Locality), and the Respective Year of the Course (UFSC, 2009)

Students come into contact with agroecology in practice in two ways. First, through courses including, *Introduction to Agroecology I* and *Agroecosystem Management I, II, III* and *IV.* Second, through the incorporation of agroecology in different forms and scales in the construction of community projects, teaching plans, and student lesson plans. The central idea is not to make teacher candidates become agroecologists per se, but to prepare them for science and EE in the context of rural education. In this case, agroecology appears more as a theme in the subjects, but also for its methodological and political potential in the practices.

In university practicum classes (and actions in schools), the challenge is to pose questions that consider the rural environment through environmental issues in dialogue with education and science education without subordinating rural knowledge. Concretely, for the past seven years, I have explored this relationship between decoloniality and agroecology through the work of supervising students in rural schools in the state of Santa Catarina. The science teaching practices that we have built at the university are guided by the fundamentals of critical environmental education, but they articulate agroecology as another place of creative opening to break with a certain consensus in school culture that science is in the highest position among all forms of knowledge. This political position does not seek to create a destructive opposition to science and its role in western society, but rather to take it as part of a process of decolonization by countering arguments used to justify differences that marginalize knowledge and people within school (curriculum).

In the curriculum of LEdC in Brazil, agroecology does not happen only from its insertion as a theme or content. It is essential to understand agroecology's emergence as a social movement and as a way of pushing against the status quo of scientific knowledges. When thinking about curricula of teacher education and rural schools this political perspective of agroecology places social and ecological justice on the educational agenda. This view of agroecology presents an opening in teacher education for what Mignolo and Walsh (2018) calls, learning to unlearn and relearn. It seeks to avoid naive postures in education, such as those that take decolonizing as a metaphor for critical pedagogy (Tuck & Yang, 2012) and the social transformation of agriculture. More than that, decolonizing is in the field of rights, land, water, and quality education in rural areas.

Agroecology also assumes a clear methodological character for rural education. It is intertwined with practices that involve science and EE (Andreoli & Borges, 2020). Therefore, supported by the curricular proposal of the LEdC movement, it is appropriate to place agroecology at the center of the practice of learning. That is, it is a great responsibility and challenge in academia, considering:

> the central role of the institution of education in the development, maintenance, and reproduction of the modern/colonial matrix of power. For this very reason, schools and universities are necessary sites of decolonial pedagogy and praxis as well. They are the sites where many of us struggle day-to-day to transgress, disrupt, and displace modern/ colonial logics, rationalities, and world-visions as the only possibilities of humanity, humanness, existence, knowledge, and thought. They are the sites, but not the only sites of course, where the doing of decoloniality, pedagogy, and praxis need to happen. (Walsh, 2017, p. 369)

Following decolonizing practices in teacher education (Martin & Pirbhai-Illich, 2016), some concepts are essential in order to take up agroecology in the school context. Understanding schools and their routines in their territories from the notion of landscapes and agroecosystems is the basis for the production of knowledge engaged with places. Agroecological food is an aggregator and tensor to destabilize material and discursive practices of power by regenerating and relocating the place of food, in the countryside, in the city, and in people's lives.

Another important element in teacher education is that agroecology, in addition to playing an organizing role in educational practices, makes the classroom (inside and outside the school) a place and source of research. Research that starts on an equal footing in relation to different local and situated knowledges (of small farmers and their children who attend schools). This movement takes us back to the notion of praxis as a key in the decolonization process. In other words, praxis is what gives movement to decoloniality,

> from action to reflection, and from reflection upon action to new action. It is reflexive and not merely reflective. It is political, critical, and theoretical, and not merely pragmatic. And it is intentional and inventive; hopeful in its inquiry, acting, and doing; and continuous in movement, contention, consciousness, and formation. Decolonial praxis is part and parcel of processes, practices, and actions of thinking and doing that endeavor to interrupt, transgress, and transcend the modern/colonial logic and frame. (Walsh, 2017, p. 368)

In the global South (especially in rural Brazil), it is in this dialogue of knowledges (Santos, 2002; Santos et al. 2008) that popular and resistance/re-existence movements (Walsh, 2017) leave important marks on agroecology to serve both science education and environmental education. This learning focuses on the interaction of systematized knowledge (sciences) with traditional and popular knowledges, those that specifically rebel against the different forms of destruction of nature and the human condition in the face of the Eurocentric model of science, technology, and art.

The dialogue of knowledges is a region that articulates the multiplicity of existences and voices. It is the "idea of the presence of epistemological multiplicities and the possibility of non-annihilating relationships" (Meneses, 2014, p. 93) coined by Santos as ecology of knowledges (2002, 2014):

> From this perspective, science constitutes an important form of knowledge – among several –, but not the only one, that is, the ecology of knowledges. As an epistemic position and denying abstract hierarchies, the ecology of knowledges (Santos, 2003, p. 747) assumes that it is possible to decolonize modern science, giving rise to a new type of relationship between scientific knowledge and other knowledges. This relational challenge entails ensuring "equality of opportunities" for different knowledge in increasingly broader epistemological disputes with the objective of maximizing the contribution of each one of them in the construction of a more democratic, fair and participatory society, where the aim is to overcome the artificial separation between the natural and the social. (Meneses, 2014)

McCune and Sánchez (2019) proposed that by centralizing agroecology in education involves territorializing the curriculum. In other words, just as thinking about agroecological farming from the territory is key for the rural context, the specificity of each agroecosystem and the curriculum and physicality of rural schools as places must reflect and impact the territory. Rural social movements in Latin America have guided the LEdC, showing more and more that territorial processes are fundamental for critical thinking in education.

As I have highlighted, during our teaching practices at the university, we take agroecology from different perspectives to think about education, above all imagining that its place in practice is mainly articulated by the territory. Working theoretically and methodologically on scientific concepts in EE from an agroecology perspective, implies understanding the social and environmental phenomena from different scales. However, it is not enough to take agroecology as a pedagogy. Knowledges (in their multiplicity) must flood the school curriculum in order to make learning a political-pedagogical act.

Agroecology—in its historical, theoretical and methodological effect on educational practices—considers that "to rebuild the world's food systems requires the joining of cultural, political, economic, social, historical and ecological processes" to replace the agribusiness model (McCune & Sánchez, 2019, p. 14). Yet, it is with this, as a starting point, that the dialogue should extend to anti-oppressive pedagogies that are so important for decoloniality from the global South (race, gender, ethnicity, and class).

In Brazil, new elements need to be considered in the analysis of agroecology in education. Beyond the notion of agroecology as a science, practice, and social movement (Norder et al., 2016), it has become a guideline for various government policies and formal education systems: this is the case of the rural education movement through the LEdC in this chapter. Unlike technical approaches that involve the training of the agroecologist, the LEdC focuses the training of a science educator with the *ethos* of an environmental educator committed to the agendas and struggles that involve agroecology in its broad spectrum. In the same line as McCune and Sánchez (2019), I defend pedagogical principles in agroecology from the Latin American context, such as:

(i) Popular education that articulates practice-theory-practice;
(ii) Horizontal relationships in teaching-learning;
(iii) Dialogue amongst ways of knowing;
(iv) Action-based, participatory, and contextualized research;
(v) Investigation of the reality (community, institutions, local movements, youth, etc.). in different places and territories.

Finally, agroecology as a pedagogy in rural teacher education implies taking it as a mediator of the themes of daily life and rural work within education. Agroecology is part of a decolonial project for rural education, in which territories are the basis for learning and apprenticeship is taken as territoriality (Meek, 2015).

A Decolonial Practice on Agroecology and Environmental and Science Education

In my practices at UFSC, one of the main challenges has been how to articulate agroecology, science and EE beyond the idea of taking agroecology only as a focus of study and application. From a decolonial perspective (Mignolo & Walsh, 2018; Meneses, 2019), I ask how this dialogic movement can change the way we see science in school. Moreover, how can future teachers modify their practices by touching on fully consolidated methodological aspects in science and environmental education. One implication is that it means avoiding teaching only as a set of techniques or just taking the environment as a physical context, considering that the latter is inherent to one of the facets of everyday life and education in rural communities.

Between 2017 and 2018, I supervised Rafael in his teaching practice at *Licenciatura em Educação do Campo*. He was born in São Paulo in the 1970s. He grew up in different peripheral communities in the largest city in Latin America (12.32 million, IBGE, 2021). He had a discontinuous school career, dropping out and returning to school many times. In his youth, following another formative path, Rafael worked in different urban movements in the city, such as those linked to culture (i.e., *capoeira*, a Brazilian martial art created by enslaved Africans in Brazil). Rafael emigrated to the island of Santa Catarina in the 1990s and became a cultural activist in the Rio Vermelho community, a suburb neighborhood of the city. In addition to working as an environmental guide in the region, he continued his experience in recent years linked to cultural activities. Through *capoeira*, he got involved in discussions about race and ethnicity, becoming a reference on this topic. Rafael had an incredible ability to provoke dialogue wherever he went. Between 2015 and 2019, he attended the LEdC at UFSC. In classes, he became a leader for colleagues and professors to bring elements of culture into discussions of rural education, especially in debate around science and environmental education. He introduced himself as an environmental educator.

During his practicum experiences, Rafael explored some elements from his background on culture and education. In 2018, he taught in an elementary school class (grade 7, 20 students, 24-hour internship divided over three months from participant-observation and teaching activities) at the State School of Basic Education Gama Rosa, in the city of São Pedro de Alcântara, Florianópolis region, 40 km away from the capital. The school is located in the city center and has approximately 615 students (SED/SC, 2021); It is considered a rural school because it predominantly serves students from rural areas (Brasil, 2010). São Pedro de Alcântara has a population of 5,935 people (IBGE, 2021). It is known as one of the first cities in Brazil to receive German and Luso-Azorean settlers in the first half of the nineteenth century (Philippi, 1995; PMSPA, 2021). In addition to being a city located in a region with a history of decimating Indigenous people (specifically, the Guarani, Xokleng and Kaingang peoples), in this same period, practices of exclusion of Afro-descendant families by European settlers occurred. A mark of this cultural issue can be seen in the construction of villages in the city (as in the case of Vila Abissínia), (Medeiros, 2019; Silva & Otto, 2010), which today house the descendants of slaves purchased by German colonizers in the post-slavery period (1888) in Brazil. The black people of this village received the right to live in the city, but they have permanently been excluded from various cultural and social activities during the last decades. During the process of organizing his practicum, Rafael deepened his studies about the place and felt challenged when touching on the theme of diversity. This whole story was central to his decisions and the way he organized his practice.

Due to his affinity to popular education, Rafael was motivated to impact the students' classroom. He explored the history of the territory as a background for science education by focusing on European colonization and its relationship to black people. One of the main challenges was how to carry out this articulation without losing the focus on problematizing science teaching and environmental education, while avoiding making his class just a contextual and historical description of the territory.

It was through a pedagogical experience with the garden and compost bin in the schoolyard that he sought to sensitize the class to the importance of diversity. The strategy was to use interculturality in a broader sense (Petri & Fonseca, 2020; Walsh, 2010). In this strategy, the garden and the compost bin were metaphorically the meeting place for stories, different ways of knowing and scientific concepts. Based on the principles of agroecology, Rafael covered themes that touched on the land struggle by movements such as the MST and Via Campesina, the local food culture and the influence of different ethnic groups (Indigenous, black and colonizers). At the same time, he questioned the place of knowledges in anti-racist education. Drawing on his experience as a *capoeirista*, he used *capoeira* as his entryway to a playful approach to children. For example, even though there is no direct relationship between *capoeira* and the concepts he would use in the classroom, he explored the *"ladainhas"* as music to mark the beginning and end of classes. The playful use of it was loaded with messages and stories of cultural appreciation of Afro-descendants in Brazil and methodologically organizing the times of classes (beginning and end).

As I discussed, one of the central issues for Rafael was how to bring his passion and commitment to diversity to the discussions during practicum. In other words, the proposal was designed based on strategies that articulated diversity and science education. Since he had practical experience with school gardens, he focused his teaching plan based on the dialogue between agroecology and environmental education. Using a more mainstream perspective of critical EE practices, he presented to the school community the theme "Garden in School: an interdisciplinary, scientific and popular space". Rafael proposed to deepen reflections on healthy eating based on what he called the "principles of agroecology" (teaching report).

Taking care not to address the concept of diversity as a certain pedagogical common sense in the educational sphere (for instance, only as an inclusion of the different/diverse), Rafael strategically avoided delving into it in the first moment of the practicum. Thus, he chose to talk much more about the place of agroecology in the science classroom and its role in the recovery of the subjects' relationships with the land and with environmental sustainability practices in rural areas. In our supervising meetings, he always stressed that he would not only like to "teach agroecology", but rather use it in its broad spectrum (science, movement and practice) as an environmental pedagogy that also cut across ethnic and cultural issues,

> I know that Agroecology is a scientific and strategic approach that corresponds to the application of concepts and principles of Ecology, Agronomy, Anthropology, and many other areas of knowledge to redesign and manage agrosystems that we want to make more sustainable throughout time. By the way, we heard this many times in the LEdC. But I see the Garden from the agroecological perspective as a popular space in the sense that I can use ecological concepts to dialogue with local culture. Through this I can work on healthy eating, but it does not mean to think about the biology of the environment or my body exclusively. I want to provoke reflections on the use of land for production of life and the many cultures. Is that why I ask the meaning of the garden at school? How can it be a popular space, from the place to the diverse, before being an interdisciplinary space, to think about Agroecology, human health and the environment? (Rafael, participant observation, notes from conversation)

Before teaching his classes to the group, Rafael also took walks to observe the different places in the community. Knocking on the doors of families' homes, visiting local businesses (small groceries, stores, bakeries), he collected materials that could be recycled and used to construct the garden and compost bin. This strategy aimed to make his practice known by the community and by the students' families. Also, it was a moment in which he even took the opportunity to invite the local people to engage in the garden project.

The notion of diversity, as I mentioned, was a central concept for Rafael's practicum, in the perspective of exploring how he could refer to it as a science concept when looking at the garden and compost bin. But, beyond that, his proposal also aimed to approach the cultural dimension from the same concept. Together, Rafael and I built a teaching proposal that sought to bridge the gap between the notion of diversity of life (environmental) and cultural diversity, through the idea of interculturality (Walsh, 2010; Mignolo & Walsh, 2018). To this end, we treated concepts theoretically used in other areas of knowledge and in different contexts as metaphors to be applied in classroom dialogue. For example, Rafael developed activities beyond what could be considered the most obvious and standard in garden practices (e.g., study of soil, plants and animals and their ecologies). He did this by using the garden as a place for the manifestation of social, cultural and environmental diversity. He asked the students, for example, for each element that the garden produces, how it is possible to think about the relationship between humans and their cultures.

A good strategy was to explain to the students that gardens produce life and food from "the hands and knowledge of all those who plant far from our homes" (Rafael, participant observation). As he commented on his class experiences he said: "it is from the generational knowledge, but also from the cultural diversity of the place and of the families that the diversity of foods we can find in our territory is expressed". Through hands-on work in the garden and teaching about growing agroecological foods Rafael talked to students about the historical relationship between colonizers and enslaved blacks in that region. In one of our supervising meetings, he said:

> My goal at school is always to analyze different types of culture. If you want, you can think of the term cultures as those of agriculture, but also ethnicities and their ecological relationships. In this way through dialogue, I like to think of different cultures and the relationship of communities with their past, which is not explained even in the textbooks that the school uses. Of course, as a rural educator, I talked about organic production, about healthy eating through our observation of the garden. That's why we did field trips to get to know other gardens in the community. People don't know that there are many immigrants in the city and that they are there too. They carry all these knowledges about their culture, their ecology, about their history. (Rafael, participant observation, notes from conversation)

Through his teaching practice he created situations that related the ways of life of organisms in soil and humans: "living organisms fight with the land, communities fight for the land to survive" (Rafael, participant observation). This argument aimed to problematize the relationship between land and garden, but that the struggle for the land of human cultures is an important part of the material survival of their lives, cultures and knowledges. Considering the origin of the students (most of them are children of small farmers from the city), the work with soil involved directly dealing with their knowledge about agriculture, food and local environmental issues. The school was the place for the construction of an intercultural and EE practice throughout the sciences.

> The practice sought to work the garden as an interdisciplinary, scientific and popular space, deepening themes with a direct and more obvious relationship with the principles of healthy eating. Of course, it was precisely from the principles of Agroecology and the importance of valuing and fighting for different types of knowledge. We proposed to take the notion of diversity as a cross-cutting concept (starting from biology and discussing issues involving politics and humanity). Assuming a very interdisciplinary facet, we are talking not only about systematized knowledge in areas such as science and environmental education. We talk about knowledge, politics and knowledge politics. In this case, the main part was to consider that knowledge is not neutral. I'm sure that the practical part and the agroecology movement is what gave us elements for this. (Rafael, participant observation, notes from conversation)

Gardens and composters are very present in science and EE practices in the Brazilian school context. Discussions around food production and its historical dimension and the environmental history of the territories are part of the discourse of critical approach to EE through popular education. Rafael's experience adds the intercultural dimension of territories, mainly when the cultural history itself was used to create tension between scientific and popular knowledges in the local context. The proposal to articulate concepts traditionally from different fields in science education (ecology and humanities, for example), even though they may dispute meanings such as diversity, provokes a movement towards decolonial learning at the center of science teaching practice at school. It is this type of dialogue in praxis that needs to be included in educational practice.

In a very concrete way, the intention of the practicum was to connect the social, cultural and environmental issues investigated in the municipality in the first two years of the course in that territory. As it is a school located in a predominantly agricultural region, it was essential to understand the relations of rural peoples with the land, in the present and the past. The decolonial posture was to link culture with the environment as a way to delink (Mignolo, 2007) from traditional practices of science and environmental education. Interculturality was both the basis and matrix for this move to delink.

In general, the pedagogical practices in rural schools in Brazil are attentive to the struggles that involve the rural and the place of knowledges of its peoples. However, both in planning and in practice, we proposed to go further. The marginal place given to rural education by urban-centric educational policy opens rural educational practice both to criticism and creation. If rural education incorporates struggles based on rural contradictions and ecological and educational issues it can become a place for proposals for the deconstruction and reconstruction of pedagogies. In this sense, agroecology, science and EE become a potent place for the practice of decolonial dialogue.

Final Considerations

Critical educator and thinker means being with and in the world. It means understanding oneself in a constant process of becoming where the "critical" is not a set postulate or an abstract of thought. Rather, it is a stance, posture and attitude, and actional standpoint in which one's own being and becoming are constitutive to the acts of thinking, imagining and intervening in transformation; that is, in the construction, creating and "walking" of a radically different world. (Walsh, 2015, p. 15)

Rural teacher education is part of the struggle for rural education. Agroecology and EE are a fundamental part of building sustainable societies (Earth Council, 1993). The challenge of transforming rural education in Brazil involves political and pedagogical learning about the positions we must take as rural and environmental educators. One of these positions is to show that urban/rural are categories that describe scales, but also identities and knowledges.

The right to train teachers based on land, environment and diversity is a priority in contemporary rural education. By focusing on agroecology, EE, and the ecology of knowledges, rural education provides pedagogical alternatives to formal education. In this context, decolonizing education happens much more by changing the terms and methods than by adding themes and content. Agroecology reposits EE by exploring local and marginalized knowledges in defence of rural territories. The LEdC curriculum is an alternative that reinvents the logic of school organization, disrupts concepts and disciplines, and transforms the experiences of subjects and communities in schools.

I have learned from LEdC students how building teaching practices that consider interculturality and everyday issues in people's lives can be revolutionary. If themes related to life and labor in the countryside are critical, the history of struggles and conflicts for the land are gaps in the practice of social, cultural and ecological education. Generational and cultural diversity are fundamental to the dialogue of knowledges and decolonizing education.

The decolonial project to pedagogy emerges from critical practices and shows that agroecology and EE are crucial to transforming our ways of thinking, doing and acting. Taking decoloniality as a process of becoming, unlearning and relearning (Walsh, 2015; Datta, 2018), as a researcher and educator, I am deeply affected by the experience of supervising practica and, of course, by sharing these experiences with the reader from a decolonial perspective.

References

Alves, M. Z., & Faleiro, W. (2019). Interdisciplinaridade na formação de professores em uma LEDOC: desafios de ensinar e aprender. *Revista Brasileira De Educação Do Campo*, 4, e5368. <https://doi.org/10.20873/uft.rbec.v4e5368>.

Andrade, F. M. R. de, Nogueira, L. P. M., Neves, L. do C., & Rodrigues, M. P. M. (2019). Educação do Campo em giro decolonial: A experiência do Tempo Comunidade na Universidade Federal Fluminense (UFF). *Revista Brasileira de Educação Do Campo*, 4, e7178. <https://doi.org/10.20873/uft.rbec.e7178>.

Andreoli, V., & Borges, M. G. (2020). Apresentação do Dossiê: Pesquisas e Práticas em Educação Ambiental e Educação do Campo. *Ambiente & Educação, 25*(2), 3–18. <https://doi.org/10.14295/ambeduc.v25i2.11874>.

Arroyo, M. G. (1982). Escola, cidadania e participação no campo. Em Aberto, ano 1, n. 9. <http://emaberto.inep.gov.br/ojs3/index.php/emaberto/article/view/1784/1523>.

Barbosa, L. P. (2017). Educação do Campo [Education for and by the countryside] as a political project in the context of the struggle for land in Brazil. *The Journal of Peasant Studies, 44*(1), 118–143. <https://doi.org/10.1080/03066150.2015.1119120>.

Brasil. (2002). Institui Diretrizes Operacionais para a Educação Básica nas Escolas do Campo. Ministério da Educação. <http://portal.mec.gov.br/index.php?option=com_docman&view=download&alias=13800-rceb001-02-pdf&category_slug=agosto-2013-pdf&Itemid=30192>.

Brasil. (2010). Decreto nº 7.352, de 4 de novembro de 2010. Dispõe sobre a política de educação do campo e o Programa Nacional de Educação na Reforma Agrária - PRONERA. Ministério da Educação. <http://portal.mec.gov.br/docman/marco-2012-pdf/10199-8-decreto-7352-de4-de-novembro-de-2010/file>.

Carvalho, I. C. M., & Frizzo, T. C. E. (2016). Environmental education in Brazil. In *Encyclopedia of educational philosophy and theory* (pp. 1–6). Cingapura: Springer.

Datta, R. (2018). Decolonizing both researcher and research and its effectiveness in Indigenous research. *Research Ethics, 14*(2), 1–24. <https://doi.org/10.1177/1747016117733296>.

dos Santos Araújo, A., & Porto, K. S. (2019). Vivências de estágio supervisionado em Ciências da Natureza em uma escola do campo: reflexão das práticas pedagógicas na formação inicial de professores da Educação do Campo. *Revista Brasileira De Educação Do Campo, 4*, e4132. <https://doi.org/10.20873/uft.rbec.v4e4132>.

Earth Council. (1993). Treaty on environmental education for sustainable societies and global responsibility. Brazil: Non-Governmental Organizations (NGO's) International, June 1992.

Farias, M. N., & Faleiro, W. (2020). Educação dos povos do campo no brasil: colonialidade/modernidade e urbanocentrismo. *Educação em Revista*, 1982–6621. <https://doi.org/10.1590/0102-4698216229>.

IBGE. (2021). São Pedro de Alcântara. <https://cidades.ibge.gov.br/brasil/sc/sao-pedro-de-alcantara/panorama>.

Kassiadou, A. (2018). Educação ambiental crítica e decolonial: Reflexões a partir do pensamento decolonial latino-americano. In A. Kassiadou, C. Sánchez, D. R. Camargo, M. A. Stortti, & R. N. Costa (Eds.), *Educação ambiental desde El Sur* (pp. 25–40). Macaé: Editora NUPEM.

Layragues, P. P., & Lima, G. G. C. (2014). The Brazilian environmental education macro-political-pedagogical trends. *Ambient. Soc., 17*(1), 23–40.

Leff, E. (2002). Agroecologia e saber ambiental. *Agroecologia e Desenvolvimento Rural Sustentável, 3*(1), 36–51.

Leite, V. de J., & Conceição, L. A. (2020). Práticas educativas de introdução a agroecologia nas escolas itinerantes do campo no Paraná. *Ambiente & Educação, 25*(2), 19–49. <https://doi.org/10.14295/ambeduc.v25i2.11529>.

Loureiro, C. F. B., & Layrargues, P. P. (2013). Ecologia política, justiça e educação ambiental crítica: Perspectivas de aliança contra-hegemônica. *Trabalho, Educação e Saúde*, Rio de Janeiro, *11*(01), 53–71.

Martin, F., & Pirbhai-Illich, F. (2016). Towards decolonising teacher education: Criticality, relationality and intercultural understanding. *Journal of Intercultural Studies, 37*(4), 355–372. <https://doi.org/10.1080/07256868.2016.1190697>.

McCune, N., & Sánchez, M. (2019). Teaching the territory: Agroecological pedagogy and popular movements. *Agric Hum Values, 36*, 595–610. <https://doi.org/10.1007/s10460-018-9853-9>.

Medeiros, H. S. (2019). A construção do "Tio Marcos" no contexto da Festa da Paz e Fraternidade: Memória, narrativas e religiosidades em São Pedro de Alcântar (SC). [Trabalho de Conclusão de Curso, Universidade Federal de Santa Catarina].

Medeiros, E. C., Moreno, G. S., & Batista, M. S. X. (2020). Territorialização nacional da Educação do Campo: Marcos históricos no Sudeste paraense. *Educação e Pesquisa, 46*, e224676. <https://doi.org/10.1590/S1678-4634202046224676>.

Meek, D. (2015). Learning as territoriality: The political ecology of education in the Brazilian landless workers' movement. *The Journal of Peasant Studies, 42*(6), 1179–1200. <https://doi.org/10.1080/03066150.2014.978299>.

Meneses, M. P. (2014). Diálogo de saberes, debates de poderes: possibilidades metodológicas para ampliar diálogos no Sul global. *Em aberto, 27*(91), 90–110. <https://doi.org/10.24109/2176-6673.emaberto.27i91.2423>.

Meneses, M. P. (2019). *Os Desafios do Sul: Traduções Interculturais e Interpoliticas entre saberes multi-locais para amplificar a descolonização da educação*. In Bruno A. P. Monteiro, Débora S. A. Dutra, Suzani Cassiani, Celso Sánchez, Roberto D.V. L. Oliveira (org.), Decolonialidades na Educação em Ciências. São Paulo: Editora Livraria da Física, 20–43.

Mignolo, W. D. (2007). Delinking. *Cultural Studies, 21*(2-3), 449–514. <https://doi.org/10.1080/09502380601162647>.

Mignolo, W. D., & Walsh, C. E. (2018). *On decoloniality: Concepts, analytics, praxis*. Durham: Duke University Press. <https://doi.org/10.1215/9780822371779>.

Molina, M. C. (2015). Expansão das licenciaturas em Educação do Campo: desafios e potencialidades. *Educar em Revista, 0*(55), 145–166. <https://doi.org/10.1590/0104-4060.39849>.

Molina, M. C. (2017). Contribuições das licenciaturas em educação do campo para as políticas de formação de educadores. *Educação & Sociedade, 38*(140), 587–609. <https://doi.org/10.1590/ES0101-73302017181170>.

Molina, M. C., & Hage, S. M. (2015). Política de formação de educadores do campo no contexto da expansão da educação superior. *Revista Educação em Questão, 51*(37), 121–46. <https://doi.org/10.21680/1981-1802.2015v51n37ID7174>.

Munarim, A. (2008). Trajetória do movimento nacional de Educação do Campo no Brasil Educação. *Revista do Centro de Educação, 33*(1), 57–72. <http://dx.doi.org/10.5902/19846444>.

Neto, L. B. (2016). *Educação rural no Brasil: Do ruralismo pedagógico ao movimento por uma educação do campo*. Uberlândia: Navegando Publicações.

Norder, L., Lamine, C., Bellon, S., & Brandenburg, A. (2016). Agroecology: Polysemy, pluralism and controversies. *Amb Soc, 13*, 1–20. <https://doi.org/10.1590/1809-4422ASOC129711V1932016>.

Oliveira, C. L. de. (2019). Os professores em formação no LE Campo: Uma análise decolonial. *@rquivo Brasileiro De Educação, 6*(15), 4–26. <https://doi.org/10.5752/P.2318-7344.2018v7n15p4-26>.

Paim, R. O. (2016). Educação Ambiental e agroecologia na educação do campo: Uma análise de sua relação com o entorno produtivo. *Revista Brasileira De Educação Ambiental (RevBEA), 11*(2), 240–262. <https://doi.org/10.34024/revbea.2016.v11.2152>.

Petri, M., & Fonseca, A. B. (2020). Entre a educação ambiental e a agroecologia:um olhar sobre Escolas Famílias Agrícolas (EFAs). *Ambiente & Educação, 25*(2), 369–392. <https://doi.org/10.14295/ambeduc.v25i2.11546>.

Philippi, A. J. (1995). *São Pedro de Alcântara: A primeira colônia alemão de Santa Catarina*. Florianópolis: Letras Brasileiras. PMSPA. (2021). História da colonização. <https://www.pmspa.sc.gov.br/cms/pagina/ver/codMapaItem/48079>.

Procampo. (2008). Edital nº 2, de 23 de abril de 2008. Chamada pública para seleção de projetos de instituições públicas de ensino superior para o Procampo. <http://portal.mec.gov.br/arquivos/pdf/edital_procampo.pdf>.

Procampo. (2009). Edital de convocação nº 09, de 29 de abril de 2009. <http://portal.mec.gov.br/dmdocuments/edital_procampo_20092.pdf>.

Pronacampo. (2012). Edital de seleção nº 02/2012-SESU/SETEC/SECADI/MEC de 31 de agosto de 2012. <http://pronacampo.mec.gov.br/images/pdf/edital_%2002_31082012.pdf>.

Quijano, A. (2000). Coloniality of power and Eurocentrism in Latin America. *International Sociology, 15*(2), 215–232. <https://doi.org/10.1177/0268580900015002005>.

Rockwell, E. (2009). *La experiencia etnográfica: historia y cultura en los procesos educativos*. Buenos Aires: Paidós.

Sánchez, C., Pelacani, B., & Accioly, I. (2020). Educação ambiental: Insurgências, reexistências e esperanças. *Ensino, Saúde e Ambiente*, Número especial, 1–20. <https://doi.org/10.22409/resa2020.v0i0.a43012>.

Santos, B. S. (2002). Para uma Sociologia das Ausências e uma Sociologia das Emergências. *Revista Crítica de Ciências Sociais, 63*, 237–280.

Santos, B. S. (2003). Para uma Sociologia das Ausências e uma Sociologia das Emergências. In B. S. Santos (Ed.), *Conhecimento prudente para uma vida decente: "Um discurso sobre as Ciências" revisitado* (pp. 735–775). Porto: Afrontamento.

Santos, B. S. (2014). *Epistemologies of the South: justice against epistemicide.* Boulder Paradigm Publishers.

Santos, B. S., Nunes, J. A., & Meneses, M. P. (2008). Introduction: Opening up the canon of knowledge and the recognition of difference. In Boaventura de Sousa Santos (Ed.), *Another knowledge is possible: Beyond Northern epistemologies* (pp. x–lxii). New York: Verso.

SED/SC. (2021). Escola de Educação Básica Gama Rosa. <http://serieweb.sed.sc.gov.br/cadueendportal.aspx?gFI6dJUHOYnHC73MPdqA40PyLoX3brxsB+LJ7mJM GKc=>.

Silva, C. T., & Otto, C. (2010). Memórias do cotidiano escolar: Encontros e desencontros entre negros e alemães em São Pedro de Alcântara, SC. Fazendo Gênero 9 - Diásporas, Diversidades, Deslocamentos. <http://www.fg2010.wwc2017.eventos.dype.com.br/resources/anais/1278187273_ARQUIVO_GENEROCINTIAEC LARICIACOMPLETO.pdf>.

Silva, M. S. (2018). O movimento da Educação do Campo no Brasil e seu diálogo com a educação popular e a pedagogia decolonial. *Cadernos de Pesquisa: Pensamento Educacional, 13*(34), 77–94. <https://seer.utp.br/index.php/a/article/view/1412/1199>.

Thiemann, F. T., Carvalho, L. M., & Oliveira, H. T. (2018). Environmental education research in Brazil. *Environmental Education Research, 24*(10), 1441–1446. <https://doi.org/10.1080/13504622.2018.1536927>.

Tuck, E., & Yang, K. W. (2012). Decolonization is not a metaphor. *Decolonization: Indigeneity, Education & Society, 1*(1), 1–40.

UFSC. (2009). Solicitação do reconhecimento e avaliação do curso de licenciatura em Educação do Campo para o MEC. <https://ledoc.paginas.ufsc.br/files/2014/07/PPP-LEdoC.pdf>.

Walsh, C. (2010). Interculturalidad crítica y educación intercultural. In J. Viaña, L. Tapia, & C. Walsh (Eds.), *Construyendo interculturalidad crítica* (pp. 75–96). La Paz, Bolivia: Instituto Internacional de Integración del Convenio Andrés Bello III-CAB.

Walsh, C. E. (2015). Decolonial pedagogies walking and asking. Notes to Paulo Freire from AbyaYala. *International Journal of Lifelong Education, 34*(1), 9–21. <https://doi.org/10.1080/02601370.2014.991522>.

Walsh, C. (2017). Decoloniality, pedagogy, and praxis. In M. A. Peters (Ed.), *Encyclopedia of educational philosophy and theory* (pp. 366–370). Singapore: Springer Nature Singapore Pte Ltd.

Wezel, A., Bellon, S., Doré, T., Francis, C., Vallod, D., & David, C. (2011). Agroecology as a science, a movement and a practice. In E. Lichtfouse, M. Hamelin, M. Navarrete, & P. Debaeke (Eds.), *Sustainable agriculture* (pp. 27–43). Dordrecht: Springer. <https://doi.org/10.1007/978-94-007-0394-0_3>.

Zeferino, J. C., dos Passos, J. C., & Paim, E. A. (2019). Decolonialidade e educação do campo: diálogos em construção. *Momento - Diálogos Em Educação, 28*(2), 21–36. <https://doi.org/10.14295/momento.v28i2.9125>.

4. *Taking Learning Outside*

Alice Johnston

Writing from Ka'tarohkwi, territory shared by the Anishinaabe and Haudenosaunee peoples.

Introduction

Within Canada and other settler colonial states, Indigenous students struggle to connect to the content and process of formal education (Battiste, 2013). This is reflected in graduation rates for Indigenous students that are often below the national average (Goulet & Goulet, 2014; Schissel & Wotherspoon, 2003). For instance, 52% of the Aboriginal population living on reserve aged twenty to twenty-four have not completed grade twelve (Richards, 2017). In comparison, only 8% of the general population from within the same age bracket have failed to finish high school (Richards, 2017). Eurocentrism, manifesting itself in cognitive imperialism, is responsible for these low levels of Indigenous student educational engagement and attainment (Battiste, 2000; Battiste, 2013; Graveline; 1998; Cajete, 1994). Indigenous students struggle to succeed in institutions where their ways of knowing, being, and relating to the world are not adequately represented. Eurocentrism also negatively impacts non-Indigenous learners, many of whom struggle to thrive emotionally, physically, spiritually, and mentally in Canada's public education system (Brendtro & Brokenleg, 2009). While attempts have been made to address these realities, few program interventions have substantively impacted Indigenous student success (Kirkness, 2013). Outdoor programming, due to its parallels with traditional Indigenous pedagogies, presents itself as potentially generative of student achievement. To better understand its potential efficacy, however, it is integral to understand the impact of colonialism on Indigenous learners.

The most commonly theorized form of colonialism is exogenous colonialism that occurred when colonizers extracted goods from colonies in order to transport them back to the imperial center. The movement of spices from India to Europe represents a historic example of this process (Tuck, McKenzie, & McCoy, 2014; Tuck & Yang, 2012). Another form of colonialism, that unfolded in North America, is settler colonialism. Because settlers never intended to return to the imperial center, settler colonialism required that Europeans justify their continued occupation of Indigenous lands (Tuck et al., 2014; Tuck & Yang, 2012). This was achieved through the construction of an artificial binary positioning reductionist Western thinking as superior to traditional Indigenous epistemologies valuing relationality, interconnectedness, and holism (Ermine, 1995; Henderson, 2000; Seawright, 2014). As subordinate, Indigenous peoples were presented as requiring the modernizing presence of Europeans. The belief in the superiority of White culture served to justify the brutality that accompanied the colonial project (Seawright, 2014). Furthermore, Eurocentrism continues to justify the oppression minority groups experience within society today (Henderson, 2000). This is because Eurocentrism functions to make the oppression of ethnic populations appear to be normative and inevitable. The subjugation of minority groups is legitimized by creating the illusion that White, Western, cultural practices represent the universal "truth" and should therefore determine what occurs within society.

It was couched within this racist belief structure that Canadian residential schools emerged as a tool of colonialism. For over a century, Indigenous parents in Canada were legally required to send their children away to residential schools. At these schools, children were beaten for speaking their language, had their cultural beliefs labeled as 'pagan', were separated from their siblings, experienced hunger, and suffered through physical and sexual abuse (Royal Commission on Aboriginal People (RCAP), 1996). Due to the presumed superiority of European culture, residential schools were intended to be assimilationist (RCAP, 1996). The central purpose of formal education was to indoctrinate Aboriginal people into a Christian, European worldview (RCAP, 1996). Despite this stated assimilationist intention, however, the actual processes implemented in schools were not conducive to integration. In residential schools, Indigenous children were given a substandard education that did not equip them for full participation in settler society (Barman, 1986). As a result of residential schools, Indigenous people and settlers alike became convinced of the inferiority of Indigenous people, because they could not assimilate, nor could they cope with living life on the land.

While the school system has evolved and is much more proficient at meeting the needs of Indigenous learners, it is still undergirded by the belief in European superiority (National Indian Brotherhood, 1972; RCAP, 1996; Truth and Reconciliation Commission (TRC), 2015). Indigenous students are expected to succeed in Eurocentric institutions in which the curriculum and teaching processes exist in stark opposition to the ways of knowing and being that infuse their cultures (Chambers, 1999). Predictably many Indigenous students respond to this reality by disconnecting from their teachers and the curriculum (Battiste, 2013). To address this issue, schools must continuously strive to decolonize their pedagogical practices and curriculum so that they better reflect the epistemologies of Indigenous learners (Battiste, 2013; Cajete, 1994; Kirkness, 2013).

The impacts of colonization extended beyond the construction of Indigenous epistemologies as inferior. Colonialism also served to separate Indigenous peoples from their land and traditional land-based practices (Palmater, 2012; Tuck & Yang, 2012). For instance, due to the impacts of residential schools, many First Nations, Métis, and Inuit youth are unfamiliar with their traditional land-based cultural practices. Furthermore, cultural dislocation from land-based practices is often compounded by deleterious environmental impacts caused by industrial development (Simpson, 2017).

Industrialism within Canada is fueled by North America's social paradigm. A dominant social paradigm is comprised of the language usages, beliefs, concepts, and values that shape the practices and institutions of a society (Bednar, 2003). Within Western culture, these elements are heavily influenced by western science. Science is the curiosity-driven pursuit of knowledge. As a result of science, important and ground-breaking discoveries have been made that have furthered understanding of the human and natural world. Also, science is often applied in ways that make the world a better place. Kimmerer (2013) writes that when scientific discoveries are "assembled by those who have a covenant of reciprocity and respect, the [discoveries] of science build a hospital, a school, a water purification plant, a wildlife refuge, an organic farm and generate knowledge on behalf of our more-than-human relatives" (p. 56). Despite this potential for positive application, however, science is often utilized within contemporary society to serve the purposes of capitalist expansion and industrial growth. This is because in the sixteenth and seventeenth-century modern science became pervaded by the Eurocentric capitalist worldview (Landstreicher, 2001). The capitalist worldview turned science into a tool of domination and knowledge into a commodity in service of material gain (Kimmerer, 2013).

The Eurocentric capitalist worldview emerged around the same time Europeans arrived in North America. At this time Europeans considered land valueless until "developed", and thus associated progress with human modification of the land. They also believed that humans were separate from and superior to nature. Due to this belief Europeans strategically created Indigenous peoples as less than human, to easily dispose of them (Seawright, 2014). The capitalist worldview emerged as one that saw land and its gifts as having the sole purpose of feeding the human desire for consumption. Cajete (1994) writes:

> During the Age of Enlightenment, Western culture broke with the ancient human 'participation mystique' as the basis for its relationship with Nature. It substituted a relationship based on objective scientific/rationalist thought that viewed the universe from a purely materialistic standpoint. Nature became a mass of dead matter ripe for manipulation and material gain. (p. 82)

Fed by the Eurocentric capitalist worldview, Bednar refers to the social paradigm which exists in the modern world as a techno-industrial paradigm and describes it as a paradigm obsessed with the free market, individual freedom, progress, and economic growth. These values have caused untold degrees of ecological destruction, as the environment has been utilized largely without criticism, solely for economic gain and human progress. Essentially modern society disconnected itself from the natural world to conquer it (Cajete, 1994).

North America's techno-industrial social paradigm is responsible for clashes with Indigenous peoples who have been given the right to hunt, trap, and fish on Canadian land through treaty obligations. For instance, at the beginning of my teaching career, I taught for three years at Dene High School, in the northern village of La Loche, Saskatchewan. La Loche and the adjacent reserve community of Clearwater River Dene Nation (CRDN) are located on the eastern shore of Lac La Loche and have a combined population of roughly 4,000 people. In 2014, in response to pollution caused by oil extraction, a group of trappers blockaded the road north of the community (Patterson, 2014). They cited the disappearance of their traditional practices as the rationale for the blockade. One of the group leaders, Don Montgrand stated:

> We've had enough! The animals are disappearing. Even the minnows are dying in the lakes. All of the chemicals they are dumping and burning in our local landfills and what they are leaving in the bush and running into the lakes. Even the people are dying of cancer and some are pretty young. We buried six in the last few months when we used to see maybe one person die of cancer in a year. (Patterson, 2014)

As a teacher living in the community, I saw the impacts of this clash first hand; Elder reports of local game meat unsafe for consumption caused many community members to purchase food from the local Northern store as opposed to engaging in traditional hunting and trapping practices. Community members' understanding of the land and land-based practices was slowly becoming irrelevant to their daily lived experience.

Addressing these issues is important because Indigenous youth who cannot envision a meaningful future through formal education or traditional land-based practices are increasingly overcome by social dislocation resulting in feelings of hopelessness (Bania, 2017). For example, incidences of self-harm, suicide, and intra-community violence are higher in Indigenous communities than the rest of Canada (Bombay, Matheson, & Anisman, 2014; Mihychuk, 2017; Webster, 2016). The year after I left Dene High School the impact of these factors was brought into devastating clarity.

On January 22, 2016, a 17-year old youth shot and killed 4 people and wounded 7 others, including 2 teachers at Dene High School.

To what extent might the school system have prevented this tragic incident? As a former staff member, I ask myself this difficult question.

Decolonization of Indigenous education must be viewed not simply as a cognitive process of revaluing Indigenous epistemologies, but also as a process that reconnects Indigenous learners to the land and land-based practices (Tuck et al., 2014). This chapter will examine the potential of outdoor programming to generate meaningful decolonization by both centring Indigenous knowledge and connecting students to the land and land-based practices.

Personal Experience and Reflections

La Loche, Saskatchewan, was a wonderful place to begin my teaching career. Approximately 30% of the world's Densuline people live in La Loche and 89% of the community speaks Denesuline (Statistics Canada, 2016). As a result, many rich Dene traditions continue to be practiced within the community. Despite the fact that almost all of the students attending La Loche Community School identify as either Dene or Métis, the majority of the teaching staff are non-Indigenous. Teachers typically come to La Loche from across Southern Canada for a short time to begin their teaching careers.

During my time teaching in La Loche, I saw many students succeed. I also, however, witnessed many students choose to disengage from their learning. Attrition rates at La Loche Community School, for example, were some of the highest in Saskatchewan. Fortunately, in my third year of teaching, I was asked to teach Outdoor Education and in so doing unwittingly discovered a way to engage my students in their learning and reconnect students to the land. Below I will share my reflections, gathered over nine months, teaching grade nine outdoor education twice a week to approximately fifteen students.

Because the school was situated in a First Nations community, I knew immediately that I wanted the Outdoor Education Program to make connections to traditional land-based cultural practices. I also knew that to deliver the program in a meaningful way I should engage community members in the process. Because many community members had experienced residential schools and/or the insidious impact of intergenerational trauma through schooling experiences, this was not easy. The effort to attract parents often involved the entire staff and took several months or in some cases even years. Fortunately, a Dene staff member connected me to a local Elder who lived on the land and who worked in schools in the area. Together, we planned a variety of outings that overlapped with student learning in Science, Language Arts, Social Studies, and Health. Not only were students learning outdoors but additionally they were learning by engaging in traditional cultural practices on the land.

Integrating land-based practices into my pedagogical approach helped me to incorporate Indigenous ways of knowing into the curriculum. For instance, on the land, the Elder taught the students about interdependence by explaining how traditionally, both living and non-living things were valued equally within Dene culture. This was in part because each element fulfilled an important role that enabled the land to function properly. I carried this concept back into the classroom by linking the concept of interdependence to the science curriculum focused on biotic and abiotic elements within ecosystems.

Adopting culture-based teachings also helped to improve the behavior of my most challenging students. For example, students who were typically disengaged from their learning listened intently to the stories shared by the community Elder and became absorbed in learning activities on the land. An outing spent rabbit snaring stands out in my mind. Due to the impacts of colonization, families in the community possessed different levels of knowledge about traditional practices. Consequently, while some of the students knew how to set and check rabbit snares and skin and cook rabbit, others were unsure of what to do. On day one of our rabbit snaring outing, I began to observe some interesting changes in one student who was typically distracted from his learning. First, he seemed motivated to ask questions and engage with the Elder. I had never witnessed this level of engagement from the student before. Once we got off the bus and began to look for spots to set snares, the student became a leader in the class by explaining to other students how to find a good spot and twist the wire in a manner that would make the snare stay. It was clear that the student possessed both cultural and ecological literacy and was in his element on the land teaching about traditional cultural practices. This revealed to me that the student was comfortable taking part in school-related activities when he had a previous cultural connection to the content. It became evident that through land-based learning, students were not only deepening their connection to Indigenous ways of knowing and to land-based practices but were also becoming more engaged in school.

While it is likely that learning about Indigenous values such as interdependence impacted my student's relationship to the land, it wasn't until I began cross-curricular planning with Social Studies that I noticed a significant change in my student's connection to the natural world. To introduce curriculum relevant to the lives of my students, I planned a cross-curricular unit that integrated Social Studies with Outdoor Education outcomes. The unit was inspired by the relationships I had developed with community family members that equipped me to interpret curricular outcomes through the lens of community priorities. For instance, while some of the families I worked with in Northern Saskatchewan profited from work in the oil and gas industry, others experienced the deleterious impact of pollution on their traditional land-based practices. Accordingly, I recognized that a unit focused on industrial development would not only align with Grade Eight Social Studies and Outdoor Education curricular outcomes but would also interest my students in their learning. Neo-colonialism is the ongoing economic, political, and cultural control that enables corporations and nation-states to continue to dominate subject groups through the exploitation of Indigenous land long after initial colonial occupation. Discussions surrounding the impacts of neo-colonial industry development were animated by observations of changes to the local fish population (Lacchin, 2015). This cross-curricular learning had a powerful impact on my students. Some of the students who had family members who profited from the oil and gas industry cited the positive economic benefits garnered by industry and remained in support of land development. On the other hand, the majority of the students became emotional discussing how neo-colonialism was impinging upon their ability to hunt, trap, and fish on land free from pollution. These students quite passionately vocalized their desire to spend more time on the land and not allow their traditional land-based lifeways to disappear. Further study would be required to see if these conversations galvanized students to take action in this regard.

Superficially, however, it appears that highlighting the impact of colonization on Indigenous lands helped to deepen the impact of land-based learning on students' connection to and care for the land.

By integrating outdoor programming into my teaching practice, I was able to contribute to decolonization. For instance, outdoor learning enabled me to meaningfully integrate Indigenous ways of knowing and being into my student's learning while connecting students to the land and land-based practices. This approach to decolonization mirrors that presented by Walter Mignolo (2011). For example, Mignolo suggests that the two choices available to the global community aren't simply rewesternization in the form of neo-liberalism, or dewesternization taking the shape of socialism/communism. Instead, he presents decolonization as a third option (Mignolo, 2011). What this third option requires, according to Mignolo, is a delinking from the temporally informed Western European epistemology that positions Westernized nations as modern and non-Industrialized nations as pre-modern. Instead, ways of knowing and being emerging from marginalized groups are positioned as viable options that exist alongside modernity as neither superior nor inferior but simply different. This work is done from the margins by actors who are epistemologically disobedient. For instance, this work can be done by actors who value and articulate knowledge that is ingrained in both the body and in local histories, and not exclusively in the mind. This implicitly challenges the Western construction of knowledge informed by the five senses as inferior, and in so doing challenges the dominance of European way of knowing and being (Mignolo, 2011).

Outdoor Learning

Outdoor learning is increasingly being taken up in formal education contexts across North America. Four approaches to outdoor learning that are commonly utilized, and that have been explained in great detail elsewhere (Webber, 2017), include outdoor education, place-based education, land-based education, and land education. In its infancy outdoor education was defined simply as: "education in, about, and for the outdoors" (Priest, 2010, p. 13). It is a form of outdoor learning that is geared primarily towards the humanistic pursuit of personal development and growth in the outdoors. Laterally, during the latter half of the twentieth century, as public concern for environmental protection mounted, outdoor education also came to be seen as a mechanism that could be used to generate an ethic of care for the natural world (Katz & Kolb, 1968; Kirk, 1975). Place-based education, formally named in the 1990s (Elfer, 2011), is a form of environmental education (EE) geared towards taking students out into local places to learn (Bowers, 1993; Greenwood, 2009; Gruenewald, 2003; Gruenewald & Smith, 2008; Sobel, 2004; Wattchow & Brown, 2011). Place-based educators are guided by the belief that the ethical, economic, political, and spiritual foundations upon which society rests will only be effectively questioned and subverted through the process of reconnecting students to their local places (Gruenewald, 2003; Sobel, 2004).

While outdoor and place-based education have been utilized successfully in Indigenous contexts, the theoretical orientations of these approaches are not explicitly geared towards integrating Indigenous knowledge into the learning process or reconnecting students to Indigenous land-based practices (Elliot & Krusekopf, 2017; Sorensen, 2008). Conversely, land-based learning, and land education are increasingly being taken up in attempts to decolonize Indigenous education. While these approaches share many similarities they also possess distinguishing features. To better understand their commonalities and differences, and how they both contributed to decolonization, I share the theoretical orientations of each approach below. I subsequently discuss how land-based learning and land education map onto my experience teaching Outdoor Education in Northern Saskatchewan.

Land-Based Education: Land-based education draws upon traditional Indigenous pedagogical approaches (Davis, Firman, Cook, & Dykun, 2015; Living Sky School Division, 2015). To better understand land-based education, therefore, it is helpful to understand traditional Indigenous forms of learning. Indigenous structures of education, utilized throughout history and in community today, are holistic and spirit driven. Additionally, and of most relevance to this chapter, Indigenous pedagogy is intimately connected to the land. For instance, Indigenous peoples view the land as a portal into the cosmological realm with valuable lessons to teach about how to conduct oneself in a good way in the world (Anuik, Battiste, & George, 2010; Styres, Haig-Brown, & Blimkie, 2013).

Land-based learning also occurs in community contexts under the guidance of community members and knowledge keepers (Davis et al., 2015; Living Sky School Division, 2015). Community relationships are seen as integral in helping individuals discern messages relayed to them through the land. For instance, in Indigenous pedagogical models, it is understood that it is the job of the family, community, and Elders to guide individuals in their learning (Battiste, 2013). Community members act as conduits helping youth to discern land-based teachings (Ermine, 1995; Simpson, 2014)

Land-based learning has the ability to disrupt colonialism's ideological premise that Eurocentric thought is superior to holistic Indigenous epistemologies emerging from the metaphysical realm. Alternatively, by connecting students to the land and its teachings, land-based learning allows students to understand and appreciate the value inherent in Indigenous knowledge (Anuik et al., 2010). Through this process, land-based learning generates a "two-worlds approach" or "two-eyed seeing" that equips students to see the world through a lens drawing on the best of both Indigenous and non-Indigenous epistemologies (Living Sky School Division, 2015; Bartlett, Marshall, & Marshall, 2012). While there is great diversity among Indigenous peoples and their distinct ways of knowing and being, there are also epistemological commonalities that unite Indigenous groups (Deloria & Wildcat, 2001; Henderson, 2000). Key overarching attributes of Indigenous epistemologies can be thought of as relationality, the interconnection between the sacred and the secular, and holism (Antoine, Mason, Mason, Palahicky, & Rodriguez de France, 2018; Cajete, 1994; Henderson, 2000).

Informed by this worldview, land-based education inspires students to develop reciprocal relationships with both human and non-human elements. Glen Coultard (2010) explains that Indigenous self-conception developed out of a place-based imaginary that:

> Demands that we conduct ourselves in accordance with certain ethico-political norms, which stresses, among other things, the importance of sharing, egalitarianism, respecting the freedom and autonomy of both individuals and groups, and recognizing the obligations that one has not only to other people but to the natural world as a whole. (p. 82)

These reciprocal relationships, once established, have the potential to help preserve the natural world and, in turn, traditional Indigenous land-based practices (Radu, House, & Pashagumskum, 2014; Taiaiake, 2005; Wildcat, McDonald, Irlbacher-Fox, & Coultard, 2014).

Land Education: Land education, which began to be theorized more substantively after 2010 (Tuck, McKenzie & McCoy, 2014), is an approach to outdoor education that intentionally names and problematizes the construction of Indigenous inferiority in addition to the settler-occupation of Indigenous land (Irlbacher-Fox, 2014; Paperson, 2014). Land education does this by putting Indigenous epistemological and ontological accounts of land at the center. This centering includes Indigenous understandings of land, Indigenous languages and practices in relation to land, and Indigenous critiques of settler colonialism (Tuck et al., 2014; Yerxa, 2014). Land education is premised on the notion that if colonization functioned by removing Indigenous peoples from their land, decolonization must require reconnecting Indigenous communities to the practices, languages, and relationships that emerge from land (Radu et al., 2014; Taiaiake, 2005; Wildcat et al., 2014).

Additionally, land education is critical of other forms of outdoor and environmental education, given their desire to sustainably re-inhabit local places replicate many of the problematic assumptions and imperatives of settler colonialism (Calderon, 2014; Tuck et al., 2014). For instance, land educators critique place-based education and a critical pedagogy of place's desire to reconnect students with the natural world (Greenwood, 2009; Gruenewald, 2003). Land educators state that this ignores Indigenous people's pre-existing connection to and rights to the same land (Tuck et al., 2014). Proponents of land education strive to decenter the goal of emplacing settlers on Indigenous land as the primary referent of success in outdoor education. Alternatively, land education is concerned with shaping a more equitable future in which Indigenous cultural lifeways are provided with the space to be practiced and to thrive (Ballantyne, 2014).

While all manifestations of land education precipitate a process through which participants learn in relationship to the land, land education can take place in a classroom setting in addition to outside on the land (Engel-Di Mauro & Carroll, 2014). For instance, while impacts of colonization might be explored outside by encouraging students to observe how colonial land-use policies impact the natural world, similar learning can also occur inside a classroom (Tuck et al., 2014). Calderon (2014), for instance, writes about how through examining how land is portrayed in textbooks land education can teach students to think critically about the impacts of colonization (Calderon, 2014).

Like land-based education, land education works to erode Eurocentrism. Land education, however, takes a distinct approach from land-based education by explicitly teaching students about how the construction of Indigenous inferiority and European superiority was used to justify the colonial project. Land educators also strive to examine with students the impact of colonialism and neo-colonialism on local landscapes and Indigenous populations. Land education theorists purport that this historical unpacking is integral since without it Indigenous epistemologies will continue to be viewed as peripheral when integrated alongside European knowledges (Calderon, 2014; Tuck et al., 2014). For instance, Calderon (2014) writes "before (Indigenous) viewpoints are included, work must be done to disrupt settler identity. Without the implementation of a rigorous land education, the inclusion of Indigenous viewpoints will continue to be marginalized because current trends of multiculturalism simply integrate material into existing curriculum frameworks" (Calderon, 2014).

Reflection

In the Canadian context, decolonization requires the integration of Indigenous knowledge into the education process and the reconnection of learners to the land. Land-based education successfully facilitates this process due to its centering of Indigenous knowledge and epistemologies. By reconnecting students to traditional Indigenous cultural practices, land-based learning connects students to land while fostering two-eyed seeing (Davis et al., 2015; Living Sky School Division, 2015). Working as a teacher in Northern Saskatchewan I facilitated land-based learning by engaging my students, under the guidance of community members, in cultural practices on the land. Through this experience, I saw first hand the positive impact land-based learning can have on decolonization, and concomitantly, on student learning. Providing my students with the opportunity to engage in land-based cultural practices not only enabled me to meaningfully integrate Indigenous ways of knowing and being into the curriculum but this integration, in turn, engaged my students in their learning.

To strengthen the decolonizing impact of land-based education, and prevent land-based education from unintentionally replicating colonial structures and practices, land education can be employed. Land education explicates how Eurocentric colonial and neo-colonial policies historically impacted and continue to shape the land. Providing participants with this perspective enables students to not only acknowledge the merits inherent in Indigenous ways of knowing and being, thereby deepening two-eyed seeing, but also to question the ethics of colonial expansion and the inevitability of neo-colonial land-use practices. Working as a classroom teacher I also witnessed the positive impact land education can have on decolonization. Specifically, I noticed how land education can impact student's feelings of connection to and care for the land. For example, when my students learned about how industrial development was impacting their ability to hunt, trap, and fish, the majority of students expressed their strong desire to eliminate industrial pollution and reconnect with their cultural land-based practices and lifeways.

Although I have spent time living in northern Canada, as a non-Indigenous woman, I cannot understand the nuanced reality of what it means to be an Indigenous person living in Canadian society. For instance, an assumption that shapes my reflection is that the preservation of traditional land-based practices and the eradication of extractivist industrial growth is a priority for Indigenous communities. In practice, however, reality is rarely so black and white. In some communities, Indigenous groups are divided on the issue of resource extraction, with some community members embracing industrial growth and its associated economic rewards (Vredenburg, 2019). For instance, Delbert Wapass, former chief of Thunderchild First Nation near Turtleford Saskatchewan explains that his vision for First Nations' economic sovereignty is complicated. Wapass explains that it "comes from being a chief, where the challenge is to provide for his people's needs—housing, health care, food, education—with limited means. And those challenges must be met in a society straddling its traditional lifestyle of hunting, gathering and agriculture and the dark side of living in contemporary Canada: unemployment, poverty, addictions, and suicide." (Vredenburg, 2019) As a result, Wapass is in support of resource extraction projects that have the potential to generate income for his community. While I observed the benefits of land-based learning and land education within my teaching context, it is possible that within communities that are largely in support of resource extraction, land-based education and land education would not have the same impact. Additionally, my observations are set within the context of Northern Saskatchewan. Pedagogical applications, such as hunting, trapping, and fishing, with powerful impacts on student learning in Northern Canada, may not be easily transferable to urban schools or educational systems in other countries. Accordingly, the results of this paper are not necessarily generalizable to all educational contexts and teachers of outdoor learning.

With these limiting variables in mind, this chapter submits that within particular contexts land-based learning and land education have the potential to positively impact Indigenous student educational engagement and relationship to the natural world. This suggests that there is utility in educators from across diverse contexts, searching for creative ways to engage students in both land-based practices and the process of unpacking the impacts of colonialism.

Conclusion

Since the last Residential School closed in 1996, many important efforts have been made to decolonize Canadian schools. Despite these efforts, Canadian schools are still overwhelmingly Eurocentric institutions. As a result, students from non-dominant cultural groups rarely see their cultural ways of knowing and being meaningfully reflected in the content and process of schooling. As a coping mechanism, these students commonly disconnect from their teachers and the curriculum (Battiste, 2013; Kanu, 2011; Kirkness, 2013). From this context, land-based learning emerges as a pedagogical alternative. By meaningfully integrating Indigenous knowledge into the school system, land-based education has the potential to disrupt the curriculum's

Eurocentric bias, and in turn engage Indigenous students in their learning (Anuik et al., 2010).

Additionally, Canada's current environmental paradigm is contributing to the disappearance of traditional Indigenous land-based practices (Anderson & Bone, 2009). This is because North America's dominant social paradigm is informed by Western science. While science is the curiosity-driven pursuit of knowledge and has resulted in many important discoveries, in the Western world the scientific process of inquiry has been co-opted by the Eurocentric capitalist worldview. Because the Eurocentric worldview glorifies progress and growth above all else, within the current environmental paradigm the natural world is seen as exploitable and valuable for its ability to provide economic reward (Aikenhead & Michel, 2010; Cajete, 1994; Kimmerer, 2013). Through the integration of Indigenous knowledge into the formal education system, land-based education has the potential to erode this culture of anthropocentrism and in so doing shape North America's environmental paradigm into a structure protective of ecosystem health and Indigenous land-based practices (Kulnieks, Longboat, & Young, 2010; Simpson, 2017).

Applying a land education lens to land-based learning also reminds us that colonization in settler colonial states is and always has been about land. Land education highlights this by suggesting that the pre-existing and ongoing cosmological and ontological relationships Indigenous people have to land should be taken up with students (Bang et al., 2014). Land education names and problematizes settler colonialism and encourages participants to acknowledge that Indigenous peoples had and continue to have rights to land. If we wish to move beyond what Tuck and Yang (2012) refer to as decolonization as a metaphor, I think it is important to incorporate this critique of colonialism into outdoor programming. For instance, Tuck and Yang (2012) write:

> Fanon told us in 1963 that decolonizing the mind is the first step, not the only step toward overthrowing colonial regimes. Yet we wonder whether another settler move to innocence is to focus on decolonizing the mind, or the cultivation of critical consciousness, as if it were the sole activity of decolonization; to allow conscientization to stand in for the more uncomfortable task of relinquishing stolen land. (p. 19)

As was stated over fifty years ago it is Indigenous culture-based education that will enable Canada to work towards meaningful decolonization. For example, in 1972, the National Indian Brotherhood (now the Assembly of First Nations) called for locally appropriate, culturally based education for Indigenous students (Morcom, 2017). The National Indian Brotherhood (1972) stated that culturally informed approaches have the potential to "give our children the knowledge to understand and be proud of themselves [in addition to] the knowledge to understand the world around them" (p. 1). Similarly, twenty years after that, the RCAP, called for culturally informed education for Indigenous learners (RCAP, 1996). Most recently, the TRC has analogously called Canadians to action. The TRC stated that school divisions must develop culturally appropriate curricula to improve education attainment levels for First Nations, Métis, and Inuit students (TRC, 2015). Land-based education informed by a land education lens offers itself as one approach educators can adopt to pay heed to these vital calls for culturally based education. Accordingly, the time for the education community to embrace decolonized outdoor programming to ensure that an adequate response to these calls to action circulating within Canada for the past half-century are implemented is long overdue.

References

Aikenhead, G., & Michel, H. (2010). *Bridging cultures: Indigenous and scientific ways of knowing nature.* Toronto, ON: Pearson Canada Inc.

Anderson, R. B., & Bone, R. M. (Eds.). (2009). *Natural resources and Aboriginal peoples in Canada: Readings, cases and commentary, 2nd edition.* Concord, ON: Captus Press.

Antoine, A., Mason, R., Mason, R., Palahicky, S., & Rodriguez de France, C. (2018). *Pulling together: A guide for curriculum developers.* Victoria, BC: BC Campus.

Anuik, J., Battiste, M., & George, P. (2010). Learning from promising programs and applications in nourishing the learning spirit. *Canadian Journal of Native Education, 33*(1), 63–82, 154–155.

Ballantyne, E. F. (2014). Dechinta bush university: Mobilizing a knowledge economy of reciprocity, resurgence and decolonization. *Indigeneity, Education and Society, 3*(3), 67–85.

Bang, M., Curley, L., Kessel, A., Marin, A., Suzukovich, E. S., & Strack, G. (2014). Muskrat theories, tobacco in the streets, and living Chicago as Indigenous land. *Environmental Education Research, 20*(1), 37–55.

Bania, M. (2017). Culture as catalyst: Preventing the criminalization of Indigenous youth. *Crime prevention Ottawa.* Retrieved from <https://www.thunderbay.ca/en/city-services/resources/Documents/Crime-Prevention/Culture-as-Catalyst-Prevention-the-Criminalization-of-Indigenous-Youth-final-report.pdf>.

Barman, J. (1986). Separate and unequal: Indian and white girls at All Hallows School, 1884–1920. In J. Barman, Y. Hebert, & D. McCaskill (Eds.), *Indian education in Canada: The legacy* (pp. 110–131). Vancouver: UBC Press.

Bartlett, C., Marshall A., & Marshall, M. (2012). Two-eyed seeing and other lessons learned within a co-learning journey of bringing together Indigenous and mainstream knowledges and ways of knowing. *Journal of Environmental Studies and Science, 2*(4), 331–340.

Battiste, M. (2000). Language and culture in modern society. *Reclaiming Indigenous voice and vision, 192*. British Columbia: UBC Press.

Battiste, M. (2013). *Decolonizing education: Nourishing the learning spirit*. Saskatoon, SK: Purich.

Bednar, B. C. (2003). *Transforming the dream: Ecologism and the shaping of an alternative American vision*. Albany: State University of New York Press.

Bombay, A., Matheson, K., & Anisman, H. (2014). *A preliminary study of student-to-student abuse in residential schools*. Retrieved from Aboriginal Healing Foundation: <http://www.ahf.ca/downloads/lateral-violence-english.pdf>.

Bowers, C. A. (1993). *Critical essays on education, modernity, and the recovery of the ecological imperative*. New York, NY: Teachers College Press.

Brendtro, L., & Brokenleg, M. (2009). *Reclaiming youth at risk: Our hope for the future*. Bloomington, IN: Solution Tree Press.

Cajete, G. (1994). *Look to the mountain: An ecology of indigenous education*. Asheville, NC: Kivaki Press.

Calderon, D. (2014). Speaking back to manifest destinies: A land education-based approach to critical curriculum inquiry. *Environmental Education Research, 20*(1), 24–36.

Chambers, C. (1999). A topography for curriculum theory. *Canadian Journal of Education, 24*(2), 137–150.

Coultard, G. (2010). Place against empire: Understanding Indigenous anti-colonialism. *Affinities: A Journal of Radical Theory, Culture, and Action, 4*(2), 79–83.

Coultard, G. (2014). *Red skin, white masks: Rejecting the colonial politics of recognition*. Minneapolis, MN: The University of Minnesota Press.

Davis, J., Firman, B., Cook, R., & Dykun, L. (2015). Land-based education success pathway Thompson community circle. *Voice Pathways to Success*. Retrieved from: <http://www.mysterynet.mb.ca/documents/general/Land%20Based%20Educatin%20Success%20Pathway%20-%20Thompson%20Community%20Circle.pdf>.

Deloria, V. J., & Wildcat, D. R. (2001). *Power and place: Indian education in America*. Golden, USA: Fulcrum Publishing.

Elfer, C. J. (2011). *Place-based education: A review of historical precedents in theory and practice*. Doctoral dissertation. Athens, GA: University of Georgia.

Elliot, E., & Krusekopf, F. (2017). Thinking outside the four walls of the classroom: A Canadian nature kindergarten. *International Journal of Early Childhood, 49*(3), 375–389.

Engel-Di Mauro, S., & Carroll, K. (2014). An African-centred approach to land education. In K. McCoy, E. Tuck, & M. McKenzie (Eds.), *Land education: Rethinking pedagogies of place from Indigenous, postcolonial, and decolonizing perspectives* (pp. 70–81). New York, NY: Routledge.

Ermine, W. (1995). Aboriginal epistemology. In M. Battiste & J. Barman (Eds.), *First nations education in Canada: The circle unfolds* (pp. 101–112). Vancouver: UBC Press.

Goulet, L. M., & Goulet, K. N. (2014). *Teaching each other: Nehinuw concepts and Indigenous pedagogies*. Vancouver: UBC Press.

Graveline, F. J. (1998). *Circle works: Transforming eurocentric consciousness*. Halifax, NS: Fernwood Publishing.

Greenwood, D. A. (2009). Place, survivance, and white remembrance: A decolonizing challenge to rural education in mobile modernity. *Journal of Research in Rural Education, 24*(1), 1–6.

Gruenewald, D. (2003). The best of both worlds: A critical pedagogy of place. *Educational Researcher, 32*, 3–12.

Gruenewald, D. A., & Smith, G. A. (2008). Creating a movement to ground learning in place. In G. Smith & D. Gruenewald (Eds.), *Place-based education in the global age: Local diversity* (pp. 345–358). New York, NY: Taylor & Francis Group.

Henderson, J. S. (2000). The context of the state of nature. In M. Battiste (Ed.), *Reclaiming Indigenous voice and vision* (pp. 11–33). Vancouver, BC: UBC Press.

Irlbacher-Fox, S. (2014). Traditional knowledge, co-existence and co-resistance. In K. McCoy, E. Tuck, & M. McKenzie (Eds.), *Land education: Rethinking pedagogies of place from Indigenous, postcolonial, and decolonizing perspectives* (pp. 145–158). New York, NY: Routledge.

Kanu, Y. (2011). *Integrating Aboriginal perspectives into the school curriculum*. Buffalo, NY: University of Toronto Press.

Katz, R., & Kolb, D. (1968). Outward Bound and education for personal growth. In F. J. Kelly & D. J. Baer (Eds.), *Outward Bound Schools as an alternative to institutionalization for adolescent delinquent boys*. Greenwich, CT: Outward Bound Inc.

Kimmerer, R. W. (2013). The fortress, the river, and the garden: A new metaphor for cultivating mutualistic relationship between scientific and traditional ecological knowledge. In A. Kulnieks, D. R. Longboat, & K. Young (Eds.), *Contemporary studies in environmental and indigenous pedagogies* (pp. 49–76). Rotterdam, SH: Sense Publishers.

Kirk, J. J. (1975). Outdoor education, conservation education: A quantum jump. *Journal of Outdoor Education, 9*(2), pp. 3–8.

Kirkness, V. (2013). *Creating space: My life and work in Indigenous education*. Winnipeg, MB: University of Manitoba Press.

Kulnieks, A., Longboat, D. R., & Young, K. (2010). Re-indigenizing curriculum: An eco-hermeneutic approach to learning. *Altern-native: An International Journal of Indigenous Peoples, 15*(24), 15–24.

Lacchin, Jacqueline M. (2015). The 'wretched of Canada': Aboriginal peoples and neo-colonialism. *Sociological Imagination: Western's Undergraduate Sociology Student Journal, 4*(1), 1–25.

Landstreicher, W. (2001). A balanced account of the world: A critical look at the scientific worldview. *The anarchist library*. Retrieved from <http://theanarchistlibrary.org/library/wolfi-landstreicher-a-balanced-account-of-the-world-a-critical-look-at-the-scientific-world-vie>.

Living Sky School Division No. 202. (2015). *Growth without limits; learning for all land-based learning program.* Retrieved from <https://saskschoolboards.ca/wp-content/uploads/pa15lssd.pdf>.

Mignolo, W. (2011). The roads to the future: Rewesternization, dewesternization, and decoloniality. *The darker side of western modernity: Global futures, decolonial options.* <https://doi.org/10.1515/9780822394501>.

Mihychuk, M. (2017). Breaking point: The suicide crisis in Indigenous communities. *Report of the Standing Committee on Indigenous and Northern Affairs,* Canada, 42, i–97.

Morcom, L. (2017). Indigenous holistic education in philosophy and practice, with wampum as a case study. *Foro de Educación, 15*(23), 121–138.

National Indian Brotherhood. (1972). *Indian control of Indian education: Policy paper presented to the Minister of Indian Affairs and Northern Development.* Ottawa: The Brotherhood.

Palmater, P. (2012, March 19). AFN election 2012: Stopping the assimilation of First Nations in its tracks [Web log post]. Retrieved from <http://rabble.ca/blogs/bloggers/pamela-palmater/2012/03/afn-election-2012-stopping-assimilation-first-nations-its-tracks>.

Paperson, L. (2014). A ghetto land pedagogy: An antidote for settler environmentalism. *Environmental Education Research, 20*(1), 115–130.

Patterson, B. (2014, November 21). Dene block road in Saskatchewan to stop oil companies [Web log post]. Retrieved from <https://canadians.org/blog/dene-block-road-saskatchewan-stop-oil-companies>.

Priest, S. (2010). Redefining outdoor education: A matter of many relationships. The Journal of Environmental Education, 17(3), 13–15. <https://doi.org/10.1080/00958964.1986.9941413>.

Radu, I., House, L. M., & Pashagumskum, E. (2014). Land, life, and knowledge in Chisasibi: Intergenerational healing in the bush. *Decolonization: Indigeneity, Education & Society, 3*(3), 86–105.

Richards, J. (2017, December). Census 2016: Where is the discussion about Indigenous education? *The Globe and Mail.* Retrieved from <https://www.theglobeandmail.com/opinion/census-2016-where-is-the-discussion-about-indigenous-education/article37313434/>.

Royal Commission on Aboriginal Peoples (RCAP). (1996). *Report of the Royal Commission on Aboriginal Peoples* (Vol. 3). Ottawa, ON: Canada Communication Group.

Schissel, B., & Wotherspoon, T. (2003). *The legacy of school for aboriginal people: Education, oppression, and emancipation.* Don Mills, ON: Oxford University Press.

Seawright, G. (2014). Settler traditions of place: Making explicit the epistemological legacy of white supremacy and settler colonialism for place-based education. *Educational Studies, 50*(6), 554–572. <https://doi.org/10.1080/00131946.2014.965938>.

Simpson, L. B. (2014). Land as pedagogy: Nishnaabeg intelligence and rebellious transformation. *Decolonization: Indigeneity, Education & Society, 3*(3), 1–25.

Simpson, L. B. (2017). *As we have always done: Indigenous freedom through radical resistance*. Minneapolis, MN: University of Minnesota Press.

Sobel, D. (2004). *Place-based education: Connecting classrooms & communities*. Great Barrington, MA: The Orion Society.

Sorenen, M. (2008). STAR: Service to all relations. In D. A. Gruenewald & G. A. Smith (Eds.), *Place-based education in the global age: Local diversity* (pp. 49–64). New York, NY: Taylor & Francis Group.

Statistics Canada. (2016). *La Loche, Northern Village [Census Subdivision], Saskatchewan and Saskatchewan [Province]*. Ottawa, ON: Statistics Canada. Retrieved August 27, 2021 from <https://www12.statcan.gc.ca/census-recensement/2016/dp-pd/prof/details/page.cfm?Lang=E&Geo1=CSD&Code1=4718074&Geo2=PR&Code2=47&SearchText=La%20Loche&SearchType=Begins&SearchPR=01&B1=All&GeoLevel=PR&GeoCode=4718074&TABID=1&type=0>.

Styres, S., Haig-Brown, C., & Blimkie, M. (2013). Towards a pedagogy of land: The urban context. *Canadian Journal of Education, 36*(2), 34–67.

Taiaiake, A. (2005). *Wasa'se: Indigenous pathways of action and freedom*. Peterborough, ON: Broadview Press.

Truth and Reconciliation Canada (TRC). (2015). *Honouring the truth, reconciling for the future: Summary of the final report of the Truth and Reconciliation Commission of Canada*. Winnipeg: Truth and Reconciliation Commission of Canada.

Tuck, E., McKenzie, M., & McCoy, K. (2014). Land education: Indigenous, postcolonial, and decolonizing perspectives on place and environmental education. *Environmental Education Research, 20*(1), 1–23.

Tuck, E., & Yang, K. W. (2012). Decolonization is not a metaphor. *Decolonization: Indigeneity, Education & Society, 1*(1), 1–40.

Vredenburg, H. (2019, May 28). Why a group of First Nations wants to own the trans mountain pipeline. *The Conversation*. Retrieved from <https://theconversation.com/why-a-group-of-first-nations-wants-to-own-the-trans-mountain-pipeline-117302>.

Wattchow, B., & Brown, M. (2011). *A pedagogy of place: Outdoor education for a changing world*. Melbourne: Monash University Publishing.

Webber, G. (2017). Intricate waters: A critical literature review of place-based education. Unpublished Master's Thesis. University of Saskatchewan, Saskatoon, SK.

Webster, P. C. (2016). Canada's Indigenous suicide crisis. *The Lancet, 387*(10037), 2492–2492.

Wildcat, M., McDonald, M., Irlbacher-Fox, S., & Coulthard, G. (2014). Learning from the land: Indigenous land-based pedagogy and decolonization. *Indigeneity, Education and Society, 3*(3), I–XV.

Yerxa, J. (2014). Gii-kaapizigemin manoomin neyaashing: A resurgence of Anishinaabeg nationhood. *Decolonization: Indigeneity, Education & Society, 3*(3), 159–166.

"Walking Gently ... Through a Cultural Lens ..." (2021)
Kylie Clarke
Descendant of the Gunditjmara, Wotjobaluk, Ngarrindjeri, & Buandig Peoples. Living on and respecting the land and waters of the Wadawurrung Peoples of the Kulin Nation.

5. Decolonizing Education for Sustainable Development

ROSEANN KERR

I live and work on the territory of the Shabot Obaadjiwan First Nation and the Ardoch Algonquin First Nation on Turtle Island. My writing in this book reflects stories and experiences in the traditional territory of the Peninsular Mayan people from contemporary Indigenous communities of Ch'ol and Tzeltal peoples, and Campesinos and Campesinas of mixed ancestry.

Introduction

The ways in which we cultivate and gather the food that nourishes our bodies has a profound effect on both ourselves and what we call the 'environment'. As such, "Agriculture should not be reduced to questions of production but should be considered a deeply ontological matter that has formed humanity's ways of being, inhabiting and surviving on the earth through millennia" (Giraldo, 2013, p. 97, author translation). In other words, agriculture is culture. It reflects both our epistemology, or what we believe to be possible, as well as our ontology, our cultural ways of interacting with ecologies to meet our material needs for survival. Although agriculture has not historically had a significant place in Western environmental education (EE) discourse (Wals & Benovot, 2017), growing contributions of industrial agriculture to global greenhouse gases and its harmful effects on local ecologies and human health (McMichael et al., 2008), have more recently forced scholars and practitioners of EE and education for sustainable development (ESD) to reckon with the issues around food and how it is produced.

Wals and Benovot (2017) outline a historical perspective of EE which shows a transition from a focus on ecological literacy and nature conservation in the late nineteenth century toward a focus on changing behaviors and policy. Starting in the 1990s, they describe a transition toward more holistic, multi-stakeholder approaches to problem solving issues of energy, poverty, food and biodiversity. In their analysis of the UN Declared Decade of Education for Sustainable Development (DESD 2005–2014) Huckle and Wals (2015) contend that the decade maintained the status quo because policy makers failed to consider issues of power, equality, politics and how the neoliberal system reduces the likelihood that citizens will adopt a sustainable livelihood. They critique ESD for reproducing existing ideas about economic development and therefore preparing people to enter a system which encourages unsustainable consumption and work (Huckle & Wals, 2015).

Wals and Benovot (2017) describe a recent trend toward Environmental and Sustainability Education which involves, "rethinking humanity's place in the world and global citizenship" with the intended impact of transitioning society toward "a more relational way of being in the world and a society based on values and structures that make sustainable living the default" (p. 406). They describe emancipatory educational approaches as being necessary in the transition toward sustainability. Emancipatory education is characterized as participatory, action-oriented, collaborative education that aims to develop, "autonomous, responsible and reflective citizens who are capable to make up their own minds and follow suitable courses of action." (p. 407). They describe this approach as promoting a diversity in values and ways of knowing including Indigenous community knowledge which has long been the basis for sustainable agriculture, passed down through generations. They claim emancipatory approaches must move "beyond the dominant anthropocentric, scientific and 'Western' materialist ways of viewing the world to include local and Indigenous perspectives" (p. 407). These perspectives are important because they are based on different ways of thinking about and enacting what it means to live well, which greatly affect ESD

This chapter will explore what could be considered an example of an emancipatory educational approach called *Campesino-a-Campesino* (CaC) pedagogy which is used in the teaching and learning of agroecology (LVC, 2017a; Machín Sosa et al., 2013). *Campesino-a-Campesino* (CaC), or peasant-to-peasant learning, is a pedagogy that developed in Mexico and Guatemala in the 1970s, to share and innovate agroecology practices among *Campesinos/as* (see Holt-Giménez, 2006 for a detailed history). CaC was responsible for the spread of agroecology throughout Latin America among small-holder peasants during the 1980s and 90s including Nicaragua, Honduras and Cuba (Holt-Giménez, 2006). This pedagogy is one among several pedagogies used today to facilitate the teaching/learning of agroecology and has been identified as an important catalyst for the spread of agroecology practices in various regions in the global South (Khadse et al., 2017; Rosset et al., 2011).

Theoretical Framing

Within the *CaC* approach, knowledge is co-generated by farmers through a process where farmers share and generate solutions to on-farm issues and continue to meet and discuss practices as they adapt them through implementation (Machín Sosa et al., 2013). Depending on the context, these solutions could be innovative practices, or originate in traditional knowledge practiced by some members of the community (Rosset et al., 2011). CaC is widely labeled as a Freirean pedagogy due to its constructivist underpinnings and horizontal teaching and learning practices (Khadse et al., 2017). Several scholars have critiqued Freirain pedagogical theory and critical pedagogy more broadly, for failing to explicitly name colonization as the oppressive force and for failing to include contextual characteristics of oppression from Indigenous perspectives, values, language and ways of life (Smith, 1999; Tuck & Yang, 2012). Based on this critique and others, I seek to use decolonization theory to examine a case of CaC pedagogy in practice in several communities in Southern Mexico. Through the exploration of the perspectives of group members, facilitators and promoters, I invite readers to consider agroecology taught through *CaC* as part of a process of decolonization of ESD.

The goal of this chapter is to show how this pedagogy has the potential to inspire decolonial action (Mignolo, 2011). As Mignolo explains, decolonial action changes the terms of the conversation through promoting alternatives to Western, universalist models for knowledge and alternative power relations (Mignolo, 2007, 2011). Decolonization requires delinking from the narratives of progress and modernity as 'natural eventualities' (Mignolo, 2011). Delinking means, not only rejecting the negative identity projected upon one's self, but also rejecting the two Western meta narratives on offer: capitalism and communism. To delink is to accept, instead, the possibility of another way of thinking and being. It means rejecting the notion of the "Third World," invented by men and institutions of the "First World," and bringing awareness of coloniality into our thinking. Delinking means creating categories of thought not derived from European political and economic theory but derived through different bodies of experience. Mignolo uses world sensing and body sensing instead of world vision because, "we all inhabit different bodies, sensibilities, memories and overall world sensing," and the term world vision, "restricted and privileged by Western epistemology, blocked the affects and the realms of the senses beyond the eyes" (p. 276). The notion of body sensing acknowledges the embodiment of knowledge through experiences and memories of histories of oppression. Border thinking is the epistemological disobedience needed to think decolonially. It involves, "dwelling and thinking in the borders of local histories confronting global designs" (p. 277). Border thinking recognizes modernity not as a natural unfolding of history, but as an exported Western concept, falsely assumed to be universal. This recognition makes room for the legitimacy of decolonial versions of modernity generated by peoples outside of the Western elite (Mignolo, 2011). In other words, border thinking allows for the rejection of the notion that industrial agriculture is the only way or the best way to raise/grow food.

If, as Homi Bhabha claims, the founding moment of modernity was constructed through colonialism, then, in order to decolonize ESD, we must question the normative master narratives of modernity (Rutherford, 1990). Modernity and progress were defined, and continue to be defined, in relation to the savage other, or the archaic past (Grosfoguel, 2011; Quijano, 2000). In this view, modernity is not a good in itself, but a master narrative behind the Western colonial project. A narrative that continues to assume that those of European descent are better, smarter, more advanced. The same narrative that has globalized the capitalist system and restructured the governments of many 'Third World' nations in exchange for debt reduction. Under the disguise of international development, modernity is pushed to 'better' the lives of those who are not as modern. Particularly relevant to *Campesino/a* populations in Mexico today was the development project's perpetuation of hierarchies constructed at the time of colonization including, the privileging of Western knowledge over other forms of knowledge, urban over rural, white over other races/ethnicities, men over women, and Western forms of pedagogy over other forms (Grosfoguel, 2011; Quijano, 2000).

Post-development theorists add to critiques of modernity and development by questioning the underlying assumption of 'development as growth' (Sachs, 1992). They advocate for the disengagement of the local from dependency on external power and emphasize conviviality, Indigenous knowledge and cultural diversity (Escobar, 1995; Esteva, 1992; Sachs, 1992). Post-development theorists emphasize that each person's view is conditional, not 'normal' or 'natural' and that labeling a country as 'underdeveloped' is a process of "naturalising the norms and historical processes of the European Self" (Ziai, 2017, p. 2551). Langdon (2013) points out that Eurocentrism, "continues to colonise ideas of progress, containing the way people can imagine a better life" (p. 389). He calls for a focus on development pedagogies that move "from patronising, colonising interventionist approaches to much more mutual process" (p. 387). With this in mind, this chapter aims to examine the perspectives of participants and facilitators involved in what might be called emancipatory pedagogies for sustainable development.

In an effort to consider the history which led to situations we strive to be emancipated from, this chapter will begin with a brief historical account of agricultural modernization, its connection to colonialism and its creation of a metabolic rift that some have identified as the source of ecological imbalance creating environmental issues of today (Clauson et al., 2015). This will provide a rationale and context for the case study of CaC pedagogy that will follow. The methodology and results of the case study will then be discussed as they relate to decolonization theory. The chapter will conclude by discussing the case's relevance to the project of decolonizing ESD.

Agricultural Modernization, Colonialism and the Metabolic Rift

The spread of industrial agriculture since World War II has led to the uncoupling of agriculture from ecosystem services– chemical inputs and fossil fuels replacing ecological functions and labor (Tomich et al., 2011). For example, instead of manure from local livestock being used to replenish soil, industrial agriculture uses purchased fertilizers made from fossil fuels. This uncoupling of agriculture from natural cycles is part of a larger ethos of extraction, which justifies industrial agricultural practices that leave soils depleted in the long term, in the name of maximum productivity and profit in the short term. Writing about the soil crisis in England in the nineteenth century, Marx called this uncoupling a metabolic rift in the nutrient cycle (Clark & Foster, 2010). Urbanization and industrialization in England at the time was being fed by farms in the countryside shipping food to the cities, which disturbed various nutrient cycles by taking the nutrients from the soil in one location but depositing the waste in another. This metabolic rift has significantly widened through the process of industrialization of agriculture where today, food, fertilizers and waste are shipped across the globe.

The widening of this rift has accompanied the expansion of industrialization. In this expansion, the production of cheap food is essential (McMichael, 2013). Today global capitalist relationships between nation states ensure migrant labor to grow local food in the North and 'cheap' food imports from the global South (McMichael, 2013). When food is affordable, the cost of wage labor can remain low, allowing the continuation of capital accumulation. For example, a period of extensive accumulation of capital in Europe in the late nineteenth century depended, not only on resource extraction from colonies, but also on the production of cheap food *in* the colonies and settler states, which allowed for the reduction of the cost of labor for manu-facturing. The push for land in the colonies to supply cheap food for export to European cities drove the displacement and elimination of Indigenous peoples in North and South America for the purpose of growing grain and cattle. In parts of Mexico, Central and South America, Indigenous peoples were forced to work on plantations producing tropical agricultural exports for the European market (McMichael, 2013).

The underlying logic that allowed for the justification of these practices was that of white superiority. Later, this logic of superiority continued in the form of charity, or white man's burden to assist those now living in poverty due to the previous policies. In the U.S. for example, settler cattle farmers sold beef to the government to feed Indigenous peoples after their main food source, buffalo, had been eliminated and they had been pushed off their land (McMichael, 2013). This is one example of how the logic of moderniza-tion through intensification of agriculture created inequalities between the colonizers and the colonized, which set up the conditions for what, in the next era, would be labeled the 'developed' and the 'underdeveloped' nations. With increasing numbers of Indigenous peoples being pushed off their lands to marginal areas, the problem of poverty was created for 'more developed' nations to solve, first through intensification of agriculture and later through food aid meant to increase food security through providing food staples.

During the mid-twentieth century, the 'Fordist' model was used to push the industrialization of agriculture toward a 'green revolution' which led to a period of intense capital accumulation for nation states in the Global South (McMichael, 2013). Widely promoted as a way to end hunger and poverty by increasing agricultural production, adoption of Green Revolution tech-niques and technologies have been encouraged in the global South since the 1960s (Astier et al., 2017). The goal of the Green Revolution was to increase production of staple crops such as maize, wheat, beans and rice by applying scientific techniques for soil fertility, breeding, and chemical pest control, developed under laboratory conditions in 'developed' countries (Chávez-Mejía et al., 2001). During this era, development became synonymous with industrialization, where food was "removed from its direct link to local ecol-ogy and culture and became an input in urban diets and industrial processing plants" (McMichael, 2000, p. 21). Ideological pressure to adopt the Green Revolution model in Mexico came from the United States with the goal of, "modernizing agricultural practices to increase the productivity of soils and labour with the objective of modernizing and industrializing societies that were considered primitive and rural" (Astier et al., 2017, p. 331). The Green Revolution can be seen as an example of a colonizing force exported from the West which aimed to replace peasant and Indigenous cultural knowledge with something better. Recognizing that our current food system is built on historical and continuing colonial relations of power (Quijano, 2000), it is important to value alternatives that have been built in the margins.

Food Sovereignty and Agroecology

The food sovereignty movement grew out of resistance to globalization, and modernization, led by La Vía Campesina (LVC), an international peasant organization that evolved from the Latin American peasant movement. LVC is a network of grassroots organizations whose mission is to, "promote small-scale sustainable agriculture, build resilient agroecosystems, [and] promote social justice and dignity" (LVC, 2017b, p. 1). Food sovereignty is defined as "the right of peoples to healthy and culturally appropriate food produced through ecologically sound and sustainable methods, and their right to define their own food and agriculture systems" (Meek & Tarlau, 2016, p. 245). Rather than focusing on peasants as victims in need of foreign or state government aid, food sovereignty flips the narrative by focusing on peasant resilience and innovation. Food sovereignty reflects a desire of those producing food to control land, water, seeds, and production decisions, which are increasingly being influenced by global market forces (Martinez Torres & Rosset, 2010). Food sovereignty in theory and action is seen as "a form of resistance to neoliberal economic development, industrial agriculture, and unbalanced trade relationships" (Carney, 2012, p. 72). Importantly, Grey and Patel (2015) explain that "food sovereignty is the continuation of anti-colonial struggles in ostensibly postcolonial contexts" (p. 433).

Over the past thirty years, agroecology has been growing as a viable alternative practice to industrial agriculture for smallholder farmers in the global South, and a way to build food sovereignty. Agroecology is a set of farming practices based on principles of ecology and Indigenous knowledge with the goals of improving soil quality and managing pests, using local resources, while reducing dependence on agricultural chemicals. Its practices include biomass recycling, cover cropping, polyculture planting, crop rotation, increasing biodiversity, and natural pest control (Gliessman, 2015). Agroecology represents a way that the metabolic rift can begin to be restored by creating closed cycle systems which rely on local sources of fertility to improve soil quality. Agroecology offers a different way, a turning away from industrial agriculture.

The principles of agroecology originated within farming and Indigenous communities in Mesoamerica as they experimented and adapted to changing conditions (Altieri & Toledo, 2011). The emergence of agroecology as a science in the 1980s and 1990s, was a result of Mexican agronomic scientists being inspired by traditional Indigenous peasant farming (Astier et al., 2017). Field work in Mexico during the 1970s documented scientific evidence of benefits of traditional peasant and Indigenous farming systems. Examples of systems observed included native corn-growing systems, such as the Mayan *milpa* where corn, beans and squash are grown together in synergistic relationship (Astier et al., 2017). Over the following decades, agroecology gained legitimacy as a scientific alternative to the Green Revolution, with contributions from studies of agroecology practices worldwide (Astier et al., 2017). The adoption of agroecology has meant that peasant farms in Latin America have reduced soil erosion, regenerated soil fertility, diversified crops, and reforested hillsides, which has translated to an increase in their food production, and elimination or reduction of their dependence on external inputs (Holt-Giménez, 2006). Several studies show agroecology's promise for increasing food production, increasing resilience to climate change events, as well as reducing the detrimental effects of agriculture on the environment (Tomich et al., 2011). For example, van der Ploeg (2010) describes peasant agroecology as co-production, "the interaction and mutual transformation of human actors and living nature" (p. 4). This is fundamentally different from the extractive way that humans relate to the earth in the industrial agriculture paradigm. In the agroecology framework, humans and the rest of nature are in interdependent nurturing relationships with each other.

Methodology

Data presented in this chapter is part of my doctoral dissertation (Kerr, 2020), which was based on 3 visits to communities in the municipality of Calakmul, Mexico in 2018 and 2019. Using critical educational ethnographic methods (Howard & Ali, 2011), this case study described how *CaC* is taken up in five communities in the municipality of Calakmul, Campeche, Mexico, supported by the NGO *Fundo Para La Paz* (FPP). These five communities represented the broader case of fifteen communities learning agroecology through CaC facilitated by FPP. Through twenty-one in-depth interviews and 4 member reflections sessions (Tracy, 2010) with group members, promoters, and FPP staff, as well as observations of four agroecology workshops, two promotor meetings, and analysis of FPP documents, a case was formed to describe socio-cultural and socio-political conditions of engagement/disengagement, the motivations of *Campesinos/as*, and pedagogical tools of *CaC* which influenced the adoption of agroecology practices. The data here represents one of the four themes discussed in my doctoral thesis (Kerr, 2020). Participant quotations were selected to illuminate the ways in which CaC pedagogy has the potential to inspire decolonial action, and to highlight local challenges that impeded this potential.

In the spirit of Rosaldo's (1989) social critic, my intention is to use my privilege as a researcher, to advocate for those whose voices are not often privileged. Because I have a white body, I can only understand conceptually what it is to be marginalized (Mignolo, 2011). As I read scholars writing from the margins, advocating for a delinking from the metanarratives of modernity as a natural progression, an inherent good, I have become skeptical of judgments of a region's level of development as highly subjective and relative, depending on which aspect of society you focus on, and with whom you speak. As a privileged white person, if I am to decolonize my thinking, it is important to consider knowledge constructed in the margins, and other ways of being and cultivating/raising food that do not submit to the logic of capital, and other examples of sustainability education that attempt to stop reinforcing colonial relations of power. An acknowledgment that different knowledges are legitimate is also an acknowledgment that different ways of life are legitimate. I also acknowledge that decolonizing EE is only one step toward decolonization, which cannot be complete without repatriation of stolen Indigenous land.

As a white foreign researcher, in the settler colonial state of Mexico, my position was that of a learner. In line with critical educational ethnography, I offered my labor in a reciprocal research relationship. FPP requested the development of a document that summarized their process and approach so they could expand CaC learning networks into more communities. Using this document, they were able to obtain funding to increase their program to include 15 more communities in the area. Member reflections (Tracy, 2010), which engaged research participants in a discussion of my initial analysis were undertaken to actively include participants in analysis and provide what Tracy (2010) calls opportunities for collaboration and reflexive elaboration which make the research process open and inclusive of various opinions and ways of knowing.

Results

In the following sections I briefly outline the case context then discuss three themes:

(a) respect for *Campesinos/as* as holders of knowledge and expertise
(b) self-sufficiency, peasant autonomy & two-eyed seeing and,
(c) paternalism and the participatory pedagogy of CaC.

Case Context

Campesino/a families in Calakmul migrated to this area from other areas of Mexico on the promise of government land grants to found communities in the 1990s. Communities were founded along the edge of a large biosphere reserve in the Mayan rainforest, the territory of Peninsular Mayan people before the caste war (1847–1901). Campesinos/as in small communities (~40–150 households) here struggle to practice subsistence farming in marginal soils, using rainfed agriculture. Challenges in this region include lack of access to water, drought, thin acidic soils, recent climate variability, and lack of Indigenous knowledge among many inhabitants of agricultural practices appropriate to the ecology of the region (Rodriguez-Solorzano, 2014; Boege & Carranza, 2009). The traditional way of *La Milpa* has been largely abandoned in the municipality after several decades of government subsidies promoting the pairing of monoculture planting with agrochemicals. Traditional varieties of corn, beans and squash are still grown, along with some hybrid varieties, but rather than planting in a traditional polyculture way (*La Milpa*), they are most often grown separately in rows using herbicides, chemical fertilizers and pesticides (*FPP*, 2016).

The food grown/raised by participating families in this study was for feeding the families first, and if there was extra, it was sold, mostly to other community members or in local markets to gain income to pay for school fees, transportation and household goods. Informal trade of goods or labor among community members was also reported as a common way to satisfy household needs. Research participants interviewed identified either as *Campesinos/as* and Indigenous (Cho'ol or Tseltal), or as *Campesinos/as* living in Indigenous communities. They expressed clear priorities of maintaining their lifestyle of subsistence farming, and life in community, rather than entry into the job market and moving to urban centers. Some community members have access to large parcels of land for growing corn, beans and squash as members of the communal land tenure structure called the *ejido*. Other community members came to their community and were granted a small plot of land (approximately 50 × 50 meters) to live in the village as *pobladores*. *Campesinos/as* often raised animals in their backyards, and had fruit trees such as mango, coconut and avocado, and some also had backyard gardens. Alongside growing food, Campesinos/as generated family income through various means including part time paid labor, sale of chilis or sheep at local markets or sale of prepared foods in the community. Participants' expressed priority was to grow, when possible, rather than purchase the food needed for family self-provisioning.

In this case the two main goals of the facilitating organization *Fondo para la Paz* were working with communities toward (a) self-sufficiency and (b) encouraging the adoption of agroecological food production practices. These two goals were interrelated in a cycle where successful transition toward agroecological practices tended to increase food production, which therefore increased community self-sufficiency. This positive impact on food production led to a positive attitude toward agroecology practices. The full results of this study are beyond the scope of this chapter. Thus, this chapter will focus on ways in which participants were able to begin building self-sufficiency and barriers that promoters and facilitators faced because of a hegemony of federal government paternalism.

FPP operated many programs, but the one in which they explicitly used CaC pedagogy focused on food grown/raised for household consumption in backyard spaces including gardens, birds, pigs and sometimes sheep, depending on the plot size. CaC pedagogy, as it was taken up in this case, began with a participatory diagnostic in each community where communities were asked to identify and assess their strengths, priorities, needs, issues and resources. Out of this process, the program was designed based on the goals of each community. Infrastructure and capacity building workshops were then offered that would allow each community to meet its goals and solve issues they had identified.

Respect for Campesinos/as As Holders of Knowledge and Expertise

Once FPP had developed relationships with community members who had interests in various topics, a *Campesino/a* from each community was asked to act in the role of a promoter guide (PG). Similar to the role of an agroecology promoter as is described in literature on CaC (Machín Sosa et al., 2013) the role of the PG was to be a leader, motivator and a guide. PGs shared knowledge and practices with their community group members, which they had learned through participating in agroecology training workshops and exchanges of experience with FPP. *Intercambios de experiencia*, or 'exchanges of experience', a recognized tool of the Campesino-a-Campesino (CaC) pedagogy, were identified as important pedagogical tools wherein groups of Campesinos/as would spend several days visiting and developing relationships with Campesinos/as using agroecological models of production (Machin Sosa et al., 2013). The experience of observing a working agroecological farm was significant in validating agroecology as a viable model (Kerr, 2020).

After training and participating in exchanges of experience, PGs were asked to form and coordinate a group of community members who had a common interest in learning more about a topic of the PGs choosing, such as raising backyard pigs. They then shared their knowledge both by giving workshops in their communities for their groups, as well as one-on-one during home visits with their group members. Importantly PGs and participants were neighbors who could walk to each other's houses and understood the culture and challenges in their community. As part of their responsibilities, PGs implemented the practices they had learned in their own backyards to validate their efficacy, and to demonstrate to their neighors that these techniques were worth putting into practice. PGs visited families in their group every 15 days to check how they were implementing the learned practices, offer guidance, advice and help when things didn't go well.

The PGs met as a group each month. At these meetings facilitators worked with PGs to collaboratively plan the goals and direction of the program. As a regular practice, PGs shared photographs of their group's activities and achievements with other PGs and FPP staff, sharing stories and challenges. PGs reported that these meetings created an important network of mutual support as they learned from each other's experiences and challenges. In this way, *CaC* pedagogy shifted the power relations away from experts in possession of Western knowledge toward fellow *Campesinos/as* and away from valuing only abstract universalized knowledge toward valuing practical contextual knowledge. This shift was also evident in the way FPP staff worked with PGs toward achieving their goals. In the first few workshops given by a PG, FFP staff act as facilitators and support for PGs, but take a backseat to PGs who are the leaders of the workshop. One of the facilitators described it as follows:

> When we feel that the person is on the way to what he/she wants to do, when we go to their workshop, we only give our opinion if we consider it prudent. If not, we let them develop, so that people see that they are the promoter anyway and have the same knowledge, and they don't have to wait for the outsider to come and tell me how to do things. (Rey, Fac. Int., 03/19)

In Rey's explanation it is clear that PGs are considered leaders in their communities, being supported toward their goals by FPP staff. The leadership role of PGs as knowledge keepers within *CaC* pedagogy sets this pedagogy apart from other ways of teaching and learning. This pedagogy values the knowledge and expertise of Campesinos/as and encourages peer-to-peer learning and collaboration.

According to FPP staff, their approach was to make proposals to *Campesinos/as* with the goal of resolving issues expressed by communities during the participatory diagnostic process. They made clear that they continue to value Campesinos/as as holders of knowledge and expertise.

> Really the people from the communities we work with they already have the knowledge of how to sow. What we do is propose new adaptations for them to have a better performance, through the specialists, let's say, it's like making new proposals to them, but because knowledge is already there. (Marla, Coord. Int., 02/19)

These proposals are an invitation to fit new knowledge within the existing frame of *Campesino/a* lifestyle rather than to replace *Campesino/a* knowledge with something else. This is an important distinction. I observed the staff members of FPP working in a way that showed respect for PGs and community members as equal partners and holders of knowledge. The role of PGs as leaders responsible for the direction of projects in their communities, sets CaC pedagogy apart from other ways of teaching and learning.

As a further example of how this case of CaC demonstrated respect for Campesinos/as as holders of knowledge and expertise relates to hiring and training of staff. As an organization, FPP practiced inclusive capacity building where participants, PGs, and staff attended exchanges of experience and courses together. This promoted a culture of inclusivity and also provided training for staff, who often did not have specialized degrees in agronomy or veterinary medicine. The coordinator made clear that hiring staff from the local area was important since they had established relationships in communities, often had skill in Indigenous languages spoken in the area and awareness of the culture and unique context of the area. This demonstrates the valuing of relationship building and local cultural knowledge over outside expertise.

Ezequías, a community technician from the local area described the humble way that staff approached their work in this program as follows:

> We always told [participants] that we don't have a career in the topic of birds specifically, or the topic of sheep. We do not know perfectly, or everything. We have certain knowledge because we have also taken courses, and if we do not know how to resolve issues in the moment, we take notes and on another occasion, we resolve it. We also always say the same to promoter guides, that they should say to the participating families when they are given a workshop, and a doubt arises, 'well, I don't know, let me go ask,' but also they are promoter guides so in some way, they have been creating experiences also upon the basis of their animals and in this way they have been becoming stronger (Ezequías, CT, Int., 02/19).

This recognition of contextual experience as building expertise shows a recognition of how *Campesinos/as* are co-creating knowledge through CaC networks. The creation of agricultural knowledge by *Campesinos/as* could be seen as an example of border thinking, where knowledge emerges from local bodies with experiences of poverty and oppression. The co-creation of knowledge is key in the development of peasant empowerment, which in turn, builds capacity for community self-sufficiency.

The form and location of *CaC* workshops is critical to understanding how this pedagogy promotes respect for Campesinos/as as holders of knowledge and expertise. Workshops and technique demonstrations were located in community, were hands-on, practical in focus and led by peers (PGs) with knowledge of context and experience implementing practices. As Freire (1970, 1973) believed, knowing required a subject's action in transforming their own reality. The hands-on nature of workshops and the implementation of practices thus combine to form the action necessary for transforming subjects' own reality and for building agroecological knowledge. This promotes fellow community members as experts who can facilitate community action without waiting for outside organizations to tell them what to do. This situates communities as central locations of experimentation and innovation, which challenges the contemporary narrative that innovation occurs mainly in scientific laboratories, led by scientific experts. CaC's respect for peasants' knowledge and capacity for innovation contributes to its capacity to generate decolonial action by locating the source of knowledge in the margins/border.

Self-Sufficiency, Peasant Autonomy, and Two-Eyed Seeing

One way in which FPP facilitated the development of community self-sufficiency through CaC was by introducing agroecological practices that increased production of food, generated small increases in income or saved money in several ways. Participants reported that the agroecological practices they had learned and implemented helped them save money on inputs such as animal feed. For example, one of the participants explained, "instead of buying commercial feed, you have the capacity to make your sprouted corn, you can grow your larvae, ... you can grow your own chicken feed without having to buy it" (Candelaria, PG., Int., 02/19). Candelaria and her group were learning ways to use what they had available to feed their backyard birds by, for example, growing maggots in manure as a source of protein for the birds, or mixing their own feed with ground corn, local plants, beans and dried eggshells.

They also reported that increases in production reduced the need to purchase food for family self-provisioning. When I asked community members if their participation had contributed to their autonomy one participant said, "Ah, yes ... I do not buy chicken, we do not buy eggs, because we have them here" (Dominga, P., Int., 02/2019). The increase in production was also a source of income from selling surplus to neighbors. For example, one member of the community earned income by cooking chicken empanadas to sell to her neighbors, another sold chicks, or chickens she had raised. These results are consistent with recent research which identified the importance of synergy between the agroecology system and gains in peasant autonomy, such as reducing external inputs, and the use of approaches that prioritize "diversity, synergy, recycling and integration" (Altieri & Toledo, 2011, p. 588) leading to the subsequent reduction in dependencies on subsidies and external inputs.

The way CaC pedagogy was facilitated by FPP in this case showed a respect for both Indigenous and Western scientific knowledge. This aligns with a two-eyed seeing approach: "To see from one eye with the strengths of Indigenous ways of knowing, and to see from the other eye with the strengths of Western ways of knowing, and to use both of these eyes together" (Hatcher et al., 2012, p. 335). In supporting community identified areas of interest, FPP offered capacity building that held both Western scientific training and traditional Indigenous knowledge as equal and complementary ways of solving local issues. Training for PGs and staff members, through workshops or exchanges of experience were led by people with both Western scientific training and traditional Indigenous knowledge. For example, training was led by ethno-veterinarians (cultural/traditional animal health care practitioners), veterinarians, agronomists with Western Scientific training in sustainable agriculture and *Campesinos/as* practicing agroecology.

One of the needs expressed by community members during the participatory diagnostic was access to knowledge around animal health. Their access to veterinarian services was limited due to geographical distance and financial cost. I observed an example of respect and equal regard for both Western scientific and Indigenous knowledge in a workshop on care and nutrition for backyard pigs. In a discussion during the workshop about animal parasites, Maria explained that she had learned through a workshop facilitated by FPP that when you use a purchased anti-parasite medication you will never kill the entire population. She explained that this meant the parasites would build up immunity to medications, so there was a need to use medication in combination with what she called an "organic method". She explained to the group the way she made a mixture of local plants and garlic and how she gave these to the pigs. This was evidence that the training respected local

Indigenous knowledge of medicinal plants, as well as scientific and was rec-
ommending these knowledges be used to complement each other. Using the
strengths of multiple ways of knowing to solve issues identified by communities constitutes a rebalancing of how these knowledges have been historically
valued from a Western perspective. This approach also empowers community
members with the knowledge they need to work toward autonomy and self-
sufficiency, representing moves toward food sovereignty through equipping
community members with the knowledge they identify as needed to control
their own food production decisions.

Paternalism and the Participatory Pedagogy of CaC

FPP's explicit long-term goal of community self-sufficiency runs against
the current of paternalism that works as a hegemonic force in Calakmul.
Community group members, PGs and FPP staff explained that some community members were not used to the participatory nature of FPP programs.
They were used to government programs in which the government came
once, built infrastructure and then left. "For example, the government does
not come and build capacity, but they send someone to do it and it's done"
(Marcela, PG, MR El Manantial, 09/19). In general, the government programs they were used to did not offer workshops. FPP, on the other hand,
built their programs on a participatory model where capacity building workshops accompany the granting of infrastructure, and participating families
are responsible for building their own infrastructure. For example, several
groups who chose to focus on backyard birds received supplies and training
to build backyard hatcheries. Generally, groups were encouraged to build
infrastructure as a team, in each other's backyards, often using a set of tools
lent by FPP.

One of the main challenges faced by promoter guides was attracting and
maintaining participants in groups. PGs explained that it was difficult to keep
participants, to convince them that it was worth investing time and energy,
that learning new things was as important as the benefit of receiving fencing,
for example. As Sebastien, the husband of PG Susana explained, they have
had a hard time maintaining participants because:

> people want the program, but the problem is, they don't attend the meetings or
> the workshops. They want a program like the government programs, where the
> government gives, just like that, for very little. At times they come in just one
> hit, there is no need to participate in meetings or workshops because there are
> none [workshops]. (Sebastian, P. Int., 02/19)

I was told that sometimes staff and promoter guides were faced with the attitude of "give me something, if you don't give me something nothing changes." When I asked what people were looking for when they were asking for 'something', I was told, they are looking for material benefits. "Many times, we go for the tangible, for questions of economy, for material benefits, no?" (Olivia, PG Int., 02/19). During member reflections in El Refugio, the daughter of one of the participants said that in general in small communities people have grown accustomed to receiving something for their participation and they don't value the learning itself. Candelaria added:

> It has been a big challenge I have had, and other PGs have had, to get people to participate, try to unite them, to organize them, it has been very complicated, because sometimes people have the visualization that, 'if I come, what are you going to give me, or what are they [FPP] going to give me?' They don't have their own willingness to go. There are others that come, but with the visualization that, 'if I go out and they [FPP] give something, I'm going to be left out', so they are stuck there watching to see, waiting for it to come to them. So it has been very challenging. (Candelaria, PG Int., 02/19)

Not all, but some community members, displayed a lack of willingness to invest their time and energy, because they did not see how they would benefit from participation. A perceived lack of material benefits was a disincentive to participate. Staff, PGs and group members related this challenge to paternalism.

> This ideology, or paternalism, it comes on behalf of the government, here in Calakmul there are many subsidies, good subsidies and ones that are not so good. For example there are many subsidies for using agrochemicals or hybrid seeds here ... and yes it generates a lot of dependence because they are accustomed to it, so when we come and say, 'well, you can produce in an agroecological way with the resources you have locally', some say, 'well, what are you going to give me?' and this is a problem. (Rey, Fac. Int., 03/19)

Through discussions with PGs and community group members, it became clear that this attitude has developed out of a dependency on a government that provides; *government as father*, on which one is dependent. In a discussion about paternalism during a member reflections session, Maria Elena said:

> People are accustomed to the idea that the government provides for people. And if they do it well, or they don't do it, there are no consequences. How do I say it, they don't check, or supervise. In comparison FPP, yes they give you things, but there is continuity, they are following a process. I think this is good because it obligates us to work. They give you things, but you have to work, you have to put it into place. (Maria Elena, P., MR. Ley de Fomento, 09/19)

This is an example of how the hegemony of paternalism reinforced passivity and works against the building of community self-sufficiency. The community technician Ezequías added:

> This is a very complicated challenge, when a family is always waiting for something, not capacity building, or personal growth, but something material, something physical, but this is difficult when this is not the objective ... .Also for participation because, the government constructs it for you, *el gobierno da* [the government gives], so it is like a shock right now ... when we arrive and we say, this is self-construction, it involves family participation, it is totally different. (Ezequías, CT. Int., 02/19)

As is demonstrated in this quotation, the hegemony of paternalism creates direct challenges to supporting self-sufficiency in this case. Marla explained that PGs played an important role in their efforts to create a counter current to this hegemony:

> The promoter guides help us because they have the understanding, they are conscious of the fact that the FPP program is not about coming and giving things, but it is about making change, changes in ideas, changes in ideology and the PGs are helping explain this idea to participants. (Marla, Coord., SMR, 09/19)

Other research by Nuijten (2004) with Campesinos/as in Western Mexico suggests that some of this disengagement could be explained by a power differential and a history of mistrust of government programs that make promises, but do not follow-through. In many regions in Mexico, government rural development is carried out by NGOs who receive contracts to carry out community projects. Marla the regional coordinator emphasized the importance of developing consistent, long-term relationships with communities because they have had many experiences in which,

> an agency receives the funds from the government, implements, the funds are gone, and they go. There is no follow-up, there has been no continuity; there have even been cases where the execution has not even finished, they have left things in the middle. And then the agency runs out of money and that's it, that's it, that's it. (Marla, F. Int., 09/19)

Given this history it is understandable that an invitation to participate might be met with an attitude of disengagement and/or mistrust of the promised benefits of participation. Staff reported that in this case, trusting relationships with communities were built slowly over several years of engagement. They shared that they underestimated the length of time necessary to establish these relationships of mutual respect.

It is evident through this case that relationships of trust and mutual respect are fundamental to the functioning of *CaC* pedagogy. CaC learning networks created in this case depended on maintaining these relationships between staff and communities, between staff and PGs, between PGs and their group members and among group members themselves. FPP's recognition that their goals require long term engagement with communities in order to build capacity for self-sufficiency, shows a deep understanding and empathy for the history and current situation for communities in Calakmul. Their way of facilitating showed a focus on participation, choice and respect for the autonomy of participants. A testament to the centrality of their goal of community self-sufficiency, FPP staff shared that they felt their work was successful when groups acted independently without needing their support and shared their stories of success with them after the fact.

FPP also expressed an interrelated challenge of an ideology exported/ imposed by government rural development programs of what constitutes a dignified life that did not align with local ecology. For example, installing flush toilets was part of the federal government's rural development strategy, but *FPP* promotes the installation of dry toilets because of the lack of water, and threat of malaria in the region. Marla explained that federal government programs

> ... sell the idea that a dignified toilet is a water toilet ... .they arrive with an idea from outside that living well is living with a cement house, when they [*Campesinos/as*] know and prefer to live in a wooden house, because it is cooler, to have a palm roof because it is cooler for this climate. We are also struggling a little against that, against those ideologies that are being imposed from a place that does not understand the context and we believe we have to implement things that can serve the region, because we know the problems of the region, we know that a water toilet is going to be a focus of infection in the house, but if we use a dry toilet correctly it can be a good solution. So we have come up against this clash of systems, ideology, and beliefs, which has cost us a lot, that is, it has been difficult. (Marla, Coord., SMR, 09/19)

Of course, the relationship between *FPP* and the government programs that serve these communities was not simple. For example, a federal government program operating in the same municipality called *Sembrando Vida,* meaning sowing life, shares *FPP*'s goals of promoting agroecological methods of food production. However, the contrast between the pedagogies of these programs is important. When I asked one group what they saw as differences between *Sembrando Vida* and the *FPP* program, Maria A. D. said of *Sembrado Vida*:

It is more strict. We have to fulfill, comply with everything, we have to sow plants, citrus trees, things like that. We have to keep it all clean ... with *FPP* there is no-one obligating you, it is more, they give you seeds to plant for your own food for your house. (Maria, A. D., MR Ley de Fomento, 09/19)

When I asked if families chose what they wanted to plant on their land as part of the *Sembrando Vida* program, Marie Elena explained that "the program comes already designed, it is already known how many of each tree they want to plant. It is not what you want to plant, it is what they say. They give the order and we carry it out ..." (Marie Elena, MR Ley de Fomento, 09/19). From this description it is clear that even though *Sembrando Vida* aligns ideologically with *FPP* on the promotion of agroecology, it continues to perpetuate the hegemony of paternalism in these communities. The fact that the program comes already designed and there is no choice in what *Campesinos/as* plant, shows a lack of respect for local knowledge and autonomy, and counters the goals of food sovereignty. As part of the *Sembrando Vida* program, participants are paid for their participation in meetings and for their compliance in the work plan set out for them, thus perpetuating the paternalistic/colonial relationship with a government on which one is dependent, and to which one is obligated.

Discussion

The process of decolonizing ESD involves questioning assumptions underlying the project of development and how it continues to push toward modernization of peasant lifestyles, continuing the processes of colonization by attempting to break the relationship with land that the peasant lifestyle cultivates. If we are to decolonize ESD, we must examine issues of power, inequality and politics inherent in content, as well as pedagogy. It is also important to consider successful examples of ESD that respect Indigenous knowledges and ways of being.

In this chapter I have attempted to show the potential for *CaC* pedagogy to be a decolonizing force. First, FPP's goal of facilitating community self-sufficiency aligns with the maintenance and flourishing of the *Campesino/a* lifestyle. The *Campesinos/as* in this case expressed clear priorities of maintaining their lifestyle of subsistence farming, and life in community, rather than entry into the job market or moving to urban centers. In this context, self-sufficiency through subsistence agriculture, rather than dependence on the state, builds food sovereignty. Supporting Campesinos/as in their chosen lifestyle of subsistence agriculture runs counter to the modernization narrative which imagines all food producers entering the market and being guided by its logic (van der Ploeg, 2018). By moving toward self-sufficiency *Campesinos/as* are rejecting the negative identity projected on them as 'poor, helpless and in need of aid', and retelling the story as one of *Campesino/a* resilience. This restorying reverses the contemporary script of modernity by revaluing the knowledge and practices of those producing food through subsistence agriculture.

Second, the promotion of agroecology can be a force for decolonizing practice, since adopting agroecological practices constitutes a turning away from industrial agriculture. As Rosset et al. (2011) write, agroecology is based on principles which are applied differently depending on the context, which means "local knowledge and ingenuity of farmers must necessarily take a front seat, as farmers cannot blindly follow pesticide and fertilizer recommendations prescribed on a recipe basis by extension agents or salesmen" (p. 168). When *Campesinos/as* choose agroecology to reproduce the subsistence lifestyle that has been their custom, they are rejecting contemporary metanarratives that industrial agriculture is the best way, or the only way to grow/raise food. Further, the increases in production experienced by those implementing agroecology become a part of decolonizing practice because these increases reduce dependence on outside actors, thus increasing food sovereignty. This, in turn, strengthens peasant autonomy and reaffirms subsistence agriculture as a viable livelihood, despite material challenges.

Most importantly, the ways in which FPP used CaC pedagogy to promote agroecology can also be understood as decolonizing. Agroecology can be taught in many ways, of which some would *not* be considered decolonizing. As Mignolo (2011) explains, "decoloniality focuses on changing the terms of the conversation and not only its content" (p. 275). In this case, CaC pedagogy changes the terms of the conversation, because it changes who has power to control the discourse and create the goals. For example, by basing its program development on community goals expressed during the initial participatory diagnostic, FPP demonstrated their respect for community knowledge and experiences. CaC pedagogy creates a structure wherein each participant is considered knowledgeable, competent and able to direct their own learning and set their own goals. Since *Campesinos/as* direct the goals of the program, CaC pedagogy relies on relationships of mutual respect and collaboration between communities and NGOs. As FPP worked with communities to facilitate learning through CaC, they collaborated with community members toward each communities' own goals. Based on community identified learning goals, they created inclusive capacity building opportunities for PGs, staff and community members which respected both Indigenous knowledge and Western scientific knowledge. Promotor guides (PGs) from each community became leaders and knowledge keepers that facilitated shared learning experiences. This case reveals PGs' role within CaC as a catalyst for community-based learning and knowledge co-creation. PGs implemented practices they learnt, adapted them to the local context, shared them with their groups and co-created knowledge with their groups as they continued to learn from their collective experiences. PGs also formed an important mutual support network for sharing knowledge and challenges that strengthened their work in their respective communities.

The juxtaposition of two programs promoting agroecology, *Sembrando Vida* and the *CaC* program show the contrast between top-down and bottom-up models of ESD. Described by community members as a ready-made program, pedagogically *Sembrando Vida* locates knowledge and expertise outside of communities, with outside experts who created the program without community input. Community member's role in *Sembrando Vida* is to comply with the program set out by the federal government. This perpetuates the hegemony of paternalism that was described by participants in this case. This hegemony reinforces a colonial relationship between Indigenous communities and the state. Paternalism created consistent challenges for PGs and FPP staff in their efforts to promote participatory programming that built community capacity for self-sufficiency. Lack of motivation to become involved in FPP programming, where material benefits were not immediately evident, was identified as a challenge in this case. FPP attempted to counter this challenge by focusing on developing long term relationships of mutual respect with PGs and community members.

Thus, this case is an important example of what Wals and Benavot (2017) call emancipatory education toward sustainable development. In this case study, CaC was shown to be participatory, action-oriented and collaborative, to embrace a two-eyed seeing approach that values both local Indigenous knowledge and scientific knowledge, and to promote sustainable livelihoods through agroecology. CaC pedagogy provides an example of how sustainable development can work to recreate power relations by changing who has the power to set goals. In doing so, other ways of living besides neoliberal models based on market logics are valued, and communities can be empowered to set their own goals for sustainable development. In this case, learning agroecology through CaC with the goal of promoting community self-sufficiency supported peasant autonomy, built community capacity for food sovereignty and opened possibilities for decolonial action in these communities.

References

Altieri, M. A., & Toledo, V. M. (2011). The agroecological revolution of Latin America: Rescuing nature, securing food sovereignty, and empowering peasants. *The Journal of Peasant Studies, 38,* 587–612. <https://doi.org/10.1080/03066 150.2011.582947>.

Astier, M., Argueta, J. Q., Orozco-Ramírez, Q., González, M. V., Morales, J., Gerritsen, P. R. W., & González-Esquivel, C. (2017). Back to the roots: Understanding current agroecological movement, science, and practice in México. *Agroecology and Sustainable Food Systems, 41,* 329–348. <http://dx.doi.org/10.1080/21683 565.2017.1287809>.

Boege, E., & Carranza, T. (2009). Convivir con la selva: agricultura sostenible campesino-indígena en el contexto de la selva en el municipio de Calakmul, Campeche. *Agricultura Sostenible Campesino-Indígena, Soberanía Alimentaria Y Equidad De Género: Seis experiencias de organizaciones indígenas y campesinas en México.* D.F. MÉXICO: PIDAASSA.

Carney, M. (2012). "Food security" and "food sovereignty": What frameworks are best suited for social equity in food systems? *Journal of Agriculture, Food Systems, and Community Development, 2*(2), 71–88. <http://dx.doi.org/10.5304/jaf scd.2012.022.004>.

Chávez-Mejía, M. C., Nava-Bernal, G., Velázquez-Beltrán, L., Nava-Bernal, Y., Mondragón-Pichardo, J., Carbajal-Esquivel, H., & Arriaga-Jordán, C. (2001). Agricultural research for development in the Mexican highlands: Collaboration between a research team and campesinos. *Mountain Research and Development, 21*(2), 113–117.

Clark, B., & Foster, J. B. (2010). Marx's ecology in the 21st century. *World Review of Political Economy, 1*(1), 142.

Clauson, R., Clark, B., & Longo, S. B. (2015). Metabolic rifts and restoration: Agricultural crisis and the potential of Cuba's organic, socialist approach to food production. *World Review of Political Economy, 6*(1), 4–32.

Escobar, A. (1995). *Encountering development: The making and unmaking of the third world.* Princeton, NJ: Princeton University Press.

Esteva, G. (1992). Development. In W. Sachs (Ed.), *The development dictionary: A guide to knowledge as power* (pp. 1–23). London & New York, N.Y: Zed Books.

FPP (*Fondo Para la Paz*). (2016). Policy brief: *Estado Actual del Manejo y Aprovechamiento de la Milpa en Calakmul.* Internal document. www.fondoparalapaz.org.

Freire, P. (1970). *Pedagogy of the oppressed* (Myra Bergman Ramos, Trans.). New York: Continuum.

Freire, P. (1973). *¿Extensión o Comunicación?: La concientización en el medio rural* (Lilian Ronzoni, Trans.). MÉXICO D.F.: Siglo Veintiuno Editores.

Giraldo, O. F. (2013). Hacia una ontología de la Agri-Cultura en perspectiva del pensamiento ambiental. *Polis (Santiago), 34,* 95–115.

Gliessman, S. (2015). *Agroecology: The ecology of sustainable food systems.* Third edition. Boca Raton: CRC Press.

Grey, S., & Patel, R. (2015). Food sovereignty as decolonization: Some contributions from Indigenous movements to food system and development politics. *Agriculture and Human Values, 32,* 431–444.

Grosfoguel, R. (2011). Decolonizing post-colonial studies and paradigms of political-economy: Transmodernity, decolonial thinking, and global coloniality. *Transmodernity: Journal of Peripheral Cultural Production of the Luso-Hispanic World, 1*(1). <https://doi.org/10.1080/09502380601162514>.

Hatcher, A., Bartlett, C., Marshall, A., & Marshall, M. (2009). Two-eyed seeing in the classroom environment: Concepts, approaches, and challenges. *Canadian Journal of Science, Mathematics and Technology Education, 9*(3), 141–153.

Holt-Giménez, E. (2006). *Campesino a campesino: Voices from Latin America's farmer to farmer movement for sustainable agriculture* (1st ed.). Oakland, CA: Food First Books.

Howard, L. C., & Ali, A. I. (2016). (Critical) educational ethnography: Methodological premise and pedagogical objectives. *New Directions in Educational Ethnography.* Published online: 20 Dec 2016; 141–163.

Huckle, J., & Wals, A. E. J. (2015). The UN decade of Education for Sustainable Development: Business as usual in the end. *Environmental Education Research, 21*(3), 491–505. <https://doi.org/10.1080/13504622.2015.1011084>.

Kerr, R. (2020). *Campesino-a-Campesino pedagogy, peasant protagonism, and the spread of agroecology: A multi-site case study* (Doctoral Dissertation, Queen's University, Kingston, Canada). Retrieved from: <http://hdl.handle.net/1974/27894>.

Khadse, A., Rosset, P. M., Morales, H., & Ferguson, B. G. (2017). Taking agroecology to scale: The zero budget natural farming peasant movement in Karnataka, India. *The Journal of Peasant Studies.* <https://doi.org/10.1080/03066150.2016.1276450>.

La Vía Campesina (LVC) (2017a). *TOOLKIT peasant agroecology schools and the peasant-to-peasant method of horizontal learning.* Retrieved from: <https://viacampes ina.org>.

La Via Campesina (LVC) (2017b). *La Via Campesina annual report 2016.* Retrieved from: <https://viacampesina.org/en/index.php/organisation-mainmenu-44/what-is-la-via-campesina-mainmenu-45>.

Langdon, J. (2013). Decolonising development studies: Reflections on critical pedagogies in action. *Canadian Journal of Development Studies / Revue canadienne d'études du développement, 34*(3), 384–399. <https://doi.org/10.1080/02255 189.2013.825205>.

Machín Sosa, B., Roque Jaime, A. M., Ávila Lozano, D. R., & Rosset, P. M. (2013). Agroecological revolution: The farmer-to-farmer movement of the ANAP in Cuba. *Asociación Nacional de Agricultores Pequeños (ANAP) and La Vía Campesina.* <https://viacampesina.org/en/agroecological-revolution-the-farmer-to-farmer-movement-of-the-anap-in-cuba/>.

Martínez Torres, M. E., & Rosset, P. M. (2010). La Vía Campesina: The birth and evolution of a transnational social movement. *The Journal of Peasant Studies, 37*(1), 149–75. <https://doi.org/10.1080/03066150903498804>.

McMichael, P. (2000). The power of food. *Agriculture and Human Values, 17*, 21–33.

McMichael, P., Powles, J. W., Butler, C. D., & Uauy, R. (2008). Food, livestock production, energy, climate change and health. *Lancet, 370*, 1253–1263. <https://doi.org/10.1016/S0140-6736(07)61256-2>.

Meek, D., & Tarlau, R. (2016). Critical food systems education (CFSE): Educating for food sovereignty. *Agroecology and Sustainable Food Systems, 40*(3), 237–260.

Mignolo, W. D. (2007). Delinking. *Cultural Studies, 21*(2–3), 449–514.

Mignolo, W. D. (2011). Geopolitics of sensing and knowing: On (de)coloniality, border thinking and epistemic disobedience. *Postcolonial Studies, 14*(3), 273–283.

Nuijten, M. (2004). Peasant 'participation', rural property and the State in Western Mexico. *Journal of Peasant Studies, 31*(2), 181–209. <https://doi.org/10.1080/0306615042000224276>.

Quijano, A. (2000). Coloniality of power, eurocentrism, and Latin America. *Nepantla: Views from South, 1*(3), 533–580.

Rodriguez-Solorzano, C. (2014). Unintended outcomes of farmers' adaptation to climate variability: Deforestation and conservation in Calakmul and Maya biosphere reserves. *Ecology and Society, 19*(2), article 5.

Rosaldo, R. (1989). *Culture and truth: The remaking of social analysis*. Beacon Press.

Rosset, P. M., Machín Sosa, B., Jaime, A. M. R., & Lozano, D. R. A. (2011). The Campesino-to-Campesino agroecology movement of ANAP in Cuba: Social process methodology in the construction of sustainable peasant agriculture and food sovereignty. *The Journal of Peasant Studies, 38*(1), 161–191.

Rutherford, J. (1990). The third space. Interview with Homi Bhabha. In Ders. (Hg.), *Identity: Community, culture, difference* (pp. 207–221). London: Lawrence and Wishart.

Sachs, W. (1992). *The development dictionary: A guide to knowledge as power*. London: Zed Books.

Smith, L. T. (1999). *Decolonizing methodologies: Research and Indigenous peoples*. London: Zed Books.

Tomich, T. P., Brodt, S., Ferris, H., Galt, R., Horwath, W. R., Kebreab, E., &Yang, L. (2011). Agroecology: A review from a global-change perspective, annual review. *Environment Resource, 36*, 193–222.

Tracy, S. J. (2010). Qualitative quality: Eight "big-tent" criteria for excellent qualitative research. *Qualitative Inquiry, 16*(10), 837–851.

Tuck, E., & Yang, K. W. (2012). Decolonization is not a metaphor. *Decolonization: Indigeneity, Education and Society, 1*(1), 1–40.

van der Ploeg, J. D. (2010). The peasantries of the twenty-first century: The commoditization debate revisited. *The Journal of Peasant Studies, 37*(1), 1–30.

van der Ploeg, J. D. (2018). Differentiation: Old controversies, new insights. *The Journal of Peasant Studies, 45*(3), 489–524. <https://doi.org/10.1080/03066 150.2017.1337748>.

Wals, A. E. J., & Benavot, A. (2017). Can we meet the sustainability challenges? The role of education and lifelong learning. *European Journal of Education, 52*(4), 404–413. <https://doi.org/10.1111/ejed.12250>.

Ziai, A. (2017). Post-development 25 years after the development dictionary. *Third World Quarterly, 38*(12), 2547–2558. <https://doi.org/10.1080/01436 597.2017.138385>.

6. Decolonizing Environmental Education in Ghana

JOHN B. ACHARIBASAM

I acknowledge that I live and work on Treaty 6 territory, the Indigenous lands of the First Nation, Métis, and Inuit

Introduction

I join the discussion on decolonizing Environmental Education (EE) by drawing on the ontology and epistemology of the Kasena ethnic group in Northern Ghana. Specifically, I report the results of a qualitative study that investigated the integration of Kasena's Traditional Ecological Knowledge (TEK) into Early Childhood Environmental Education (ECEE) in Boania Primary School, Ghana. The chapter focuses particularly on the ontological and epistemological implications of Integrating TEK into ECEE for children as well as for decolonizing EE more broadly. The aim is to show how EE can be more inclusive of all voices.

In the field of Early Childhood Environmental Education (ECEE), calls have been made to incorporate local culture and Indigenous Knowledges (IKs) into programs (see Nelson, Pacini-Ketchabaw, & Nxumalo, 2018; North American Association of Environmental Education [NAAEE], 2010; Nxumalo, 2019; Nxumalo & Cedillo, 2017; Ritchie, 2012). Generally, the reasons behind these calls range from "philosophical and ethical to the utilitarian and pragmatic" (Pence & Shafer, 2006, p. 2). But the most important argument has been that IKs will help change the way children relate to nature by unsettling the Dominant Western (DW) "paradigm that views humans as superior to and separate from the more-than-human world ..." (Nxumalo & Villanueva, 2019, p. 41). Studies show ECEE aims at developing children's love and respect towards nature/environment (Born, 2018; Willson,

1994; Tilbury, 1994; NAAEE, 2010) and IKs, especially TEKs, have the cultural framework of love, respect, reciprocity, and responsibility towards nature (see Kimmerer, 2002; Reid, Teamey, & Dillon, 2002). Hence, when integrated into ECEE, TEKs can significantly help develop in children the love and respect towards nature. Based on this, Kimmerer (2012) observed in the context of Canada that TEK "builds capacity for students in regaining a relationship with ecological systems which is based on indigenous principles of respect, responsibility, and reciprocity" (p. 319). Nxumalo and Villanueva (2019) also observed that situating IKs in ECEE helps unsettle extractive and human-centric relations with nature whilst nurturing reciprocal relationships with nature instead.

In the Ghanaian context, ECEE also lacks Indigenous worldviews. EE (under which ECEE falls) was introduced as an integrated program focusing on both environment and science to the neglect of IK. According to the Global Environmental Education Partnership (2019), the Ministry of Education through the Ghana Education Service (the agency in charge of education) incorporates EE into formal education through the Integrated Science curriculum for all levels of education (Global Environmental Education Partnership, 2019). EE is taught this way from basic school to Senior High School. But the adoption of science as an approach to integrating EE into classroom topics has resulted in too much focus on Western knowledge/science to the neglect of TEK. The section below demonstrates how TEK was integrated into ECEE in a rural primary school in Northern Ghana using a Two-eyed seeing Indigenous methodology. Similarly, teachings on ECEE both under the Western eye (teacher's teachings) and the Indigenous eye (Elders' teachings) are presented to further highlight the pedagogical insufficiencies that emerge when we rely on Western science alone to deliver EE.

The Study

The study set out to investigate the integration of TEK into ECEE in a Kindergarten 2 (KG2) class. The ages of children in KG2 ranged from six to eight years. The plan for teaching and the research analysis adopted a two eyed-seeing Indigenous approach because this methodology gives equal importance to both Western knowledge and Indigenous knowledges (see; Bartlett et al., 2012; Iwama et al., 2009; Marsh, Cote-Meek, Toulouse, Najavits, & Young, 2015). As currently practiced, ECEE is situated in the subject area of science with limited Indigenous worldviews.

Data were collected by using participant observation and in-depth interviews. A total of twelve research participants were selected from

Boania Primary School and the community. This included: two Elders, one Kindergarten 2 (KG2) teacher, and nine (9) pupils from the school (KG2). First, the researcher interviewed the two community Elders and the KG2 teacher on what they knew about Kasena's TEK and how they thought it could be integrated into the ECEE curriculum and pedagogy. Second, the TEK was integrated into a KG2 curriculum (mostly environmental studies topics). These topics included: living and non-living things (including animals, domestic and wild); water, air, plants, gardening (including types of soil and gardening, making the soil fertile for gardening); light—day and night (including natural and man–made sources of light); changing weather conditions (including changing weather conditions, positive and negative effects of weather conditions); and "my local community" (see Ministry of Education, 2019).

This was done with the help of the two Elders (a man and a woman) as holders of the TEK. Kim and Dionne (2014) suggested that when integrating IK into education, it is essential to create a venue for true experts to share their knowledge directly with learners. The Elders were invited to the school once a week until the end of the study period from January 2020 to March 2020, to teach the children TEK (through outdoor learning activities) based on the environmental studies topics listed above. They alternated their visits: if a male Elder visited this week, a female Elder visited the other week to teach TEK and take the children outdoors. The researcher was present on all these visits and outdoor activities to observe. The teacher on the other hand taught these topics, usually in the classroom, as they appeared in the curriculum. The teacher taught curricular science content (environmental science subjects) once a week. This was either a day or hours before the Elders' visit to the school and this took place in the classroom. Hence, the two eyes were equally represented in line with the two-eyed seeing methodology adopted. The teacher represents the Western eye in this study since she drew on textbooks and curricular materials which were based predominantly on Western science whilst the two traditional Elders represent the Indigenous eye.

As a methodology, two-eyed seeing grew out of the teachings of the late chief Charles Labrador of the Acadia First Nation in Nova Scotia, Canada (Greenwood et al., 2015, p. 17). Acadian First Nation scholars, Albert and Murdena Marshall, (Bartlett et al., 2012) formalized the methodology for educational purposes. The methodology is currently seen as a "guiding principle for walking in two worlds" (Greenwood et al., 2015, p. 17). The main argument of the two-eyed seeing methodology is that IKs and Dominant Western-Knowledges (DW-K)can co-exist in an educational setting and children learn to see from one eye with the strengths of IKs and from the

other with the strengths of DW-K (Bartlett et al., 2012). With this methodology, none of the eyes is more important than the other. Thus, preventing
the power dominance that often emerges when IK and DW-K are brought
together. Students are encouraged to use both eyes together for the benefit
of all.

Learning Activity Under the Topic Living and Nonliving Things: February 16th to February 22nd, 2020 and February 23rd to February 29th, 2020

This section goes into detail presenting teachings of the Western eye and the
Indigenous eye under the same topic of living and non-living things.

Teacher

The teacher taught the topic of living and non-living things a day before the
Elder visited the class. The teacher and her pupils examined the dichotomy
between living and non-living things. They talked about different types of
things around them and classified them into living things and non-living
things. The teacher brought posters/pictures of living (different animals)
and non-living (stones, bags, spoons, tables, and buildings) things to the
class. The children were taught the characteristics of living things vis-a-vis
non-living things. The teacher stated that all living things grow, reproduce,
move, and eat, whilst non-living things do not grow, move, reproduce, and
eat. Additionally, children were made to draw and name examples of living
things such as animals (domestic and wild), whilst they named and drew
examples of non-living things such as stones and buildings. They also formed
sentences using examples of living and non-living things. There was no outdoor activity, but teaching was in children's mother tongue, Kasem, and in
English. The children were graded on their performances (drawings) and
all this teaching took place in the class. Teaching was more structured, and
children raised hands before they answered questions. I observed the presen-
tation by the teacher on the characteristics of living things focused more on
animals than other forms of living things (IE: there were no plants or inver-
tebrate animals in her examples). The teacher's teachings focused predominantly on cognitive development, lacked outdoor activities, and were limited
in helping children form relational, reciprocal, and respectful relationships
with nature. Concerning the lack of outdoor activities, I observed most of
the teacher's teachings of ECEE topics occurred in the classroom. Elvstam
and Fleischer (2018) observed this to be a national issue as outdoor learning
pedagogy is not largely adopted in Ghana. However, the NAAEE (2010)

argued that, for ECEE to be effective, there was a need for frequent opportunities to explore, observe, and play in natural environments. According to McCarter and Gavin (2011), most environmental knowledge is acquired at an early age through sustained contact with nature, tutelage by parents, and play with peers. Therefore, formal education without outdoor experiences has the potential to inhibit children's learning of environmental knowledge (McCarter & Gavin, 2011).

Elder (Man)

This week focused on identifying plants in the community, their traditional names and uses. Learning was more casual, playful, and children did not raise hands before speaking. According to the Elder, trees were among the most sharing and caring species in Boania because they gave things (fruits, local dye, medicine, and so on) freely to the community. The trees give these things because they care. He gave examples of some trees (*Anogeissus leiocarpus*) that are used to dye locally made bags. He included medicinal plants such as the guava, mango, tamarind tree, *Khaya senegalensis*, and neem tree. For example, leaves from the neem tree are used to treat malaria. During the outdoor learning activity, we realized that the outer bark of most of the trees (Tamarind tree, *Khaya senegalensis*) had been peeled off (harvested) for medicinal purposes. Delvaux, Sinsin, Darchambeau, and Van Damme (2009) observed that after bark harvesting, different tree species have different recovery rates. According to the scholars, two species, *Khaya senegalensis* and *Lannea kerstingii*, showed complete wound recovery whilst *Afzelia africana*, *Burkea africana*, and *Maranthes polyandra* have a very poor recovery. Hence different species are impacted differently. It is therefore important to include TEK so people learn to harvest honorably so the trees can recover.

Besides the tangible objects the community got from trees, the Elder also talked about humans learning from other forms of creation on how to live. The baobab and tamarind trees for example are often regarded as the ones who have seen it all by the Kasena because these trees live longer. The Elder concluded that the history of the community can be learnt from the baobab or tamarind tree. According to him, these trees have seen generations come and go including their ancestors. As a result, trees like these know their ancestors personally, their names, and have seen their faces. Hence, they cannot be treated as just trees but rather as part of the family and community. Scholars (see Patrut et al., 2007; Swart, 1963) have observed that the baobab tree, for example, can live for over 1,000 years. Therefore, trees like these are considered sacred and part of the community.

From the jackalberry tree which mostly grows on termite mounds, he taught the children the concept of live and let live. According to the Elder, there exists a symbiotic relationship between the jackalberry tree and the termites, neither harming the other. The roots of the tree provide a habitat for the termites and they, in turn, do not eat the roots. He says humans can emulate this to coexist with nature. Furthermore, from the *Faidherbia albida* the Elder indicated that humans must learn to use resources sustainably during bumper harvest. The tree, *Faidherbia albida*, has a reversed or inverted phenology, in that "the species has the unique characteristic of shedding its foliage at the start of the rainy season, and of coming into leaf in the dry season" (Wood, 1992, p. 9). According to the Elder, the tree saves its water during the rainy seasons when there is plenty and uses it during the dry season when there is a shortage. Based on this, the Elder stated that humans should also learn to save food when there is a bumper harvest in the rainy season to cater for food shortages during the dry season. Since agriculture is rain-fed in the community, symbolically, dry season indicates lean season (food shortage) in the community of Boania, whilst rainy season indicates time of bumper harvest because that is when farming is done. However, botanists (see Wood, 1992) rather see the reverse phenology of this tree species as making it ideal for agroforestry since it does not interfere with agriculture. As Wood (1992) noted, "This unexpected inverted phenology means that its presence in farmers' fields does not interfere with agriculture, and, indeed, makes it an ideal agroforestry tree for use in combination with crops" (p. 9). These teachings confirm studies by others that TEK has a cultural framework of respect, reciprocity, and responsibility towards nature (see Kim, Asghar, & Jordan, 2017; Parrota & Trosper, 2012; Reid et al., 2002), whereas Western science is more oriented towards trees as resources. The human centered pedagogy of viewing nature as a resource is not an effective approach to delivering EE at the early learning level (Nxumalo & Cedillo, 2017; Nxumalo, 2019).

Another important observation that emerged from the teachings on the *Faidherbia albida* was epistemic differences between TEK and Western science and the similarities in knowledge creation under the two worldviews. There were differences in knowledge held by both eyes concerning the *Faidherbia albida*. But at the same time, I observed both knowledges of the tree were acquired through a similar method of observation.

Elder (Woman)

The Elder took time to explain the day's activities to the children. The day involved two activities. First, this outdoor learning activity was a continuation of the previous week. Children were given the assignment to find out from their parents the type of trees that were forbidden to be used for fuel-wood in their various homes. Hence the first outdoor activity involved them identifying these trees not used for fuelwood. From this activity, I observed that each household tabooed different species of trees. Among the Kasena, each household or even family has a personal deity they worship. Part of this worship normally involves reverence for a certain animal or plant species. It is forbidden to either kill or burn such trees for fuelwood.

The second part of the day's activity was learning how to make traditional mats, bags, and baskets from elephant grass and straws. For this, the Elder brought along a needle and thread (made from bast fibre) to the class. Next, the class went outdoors to identify elephant grass and straws for making these items. I observed that some of the children knew how local mats, baskets, and bags were made from straws and elephant grass because they had seen them being made at home. The Elder constantly reminded the children these (elephant grass and straws) were gifts from nature. Hence, they had to be pre-served and treated with care. She talked about how in the past these grasses grew abundantly in the community. But now they are becoming scarce due to bushfires. She talked about the cultural and spiritual importance of these items especially, the mats without which the Kasena cannot bury their dead. The Elders' teachings were more casual, practical, playful, less structured, and gave children the freedom to explore nature at their own pace.

All the Elders' teachings took place outdoors. As Kimmerer (2012) con-cluded, "in my experience, the classroom is not the most conducive environ-ment for engaging TEK in environmental science, as it is far removed from the sources of traditional knowledge" (p. 320).

Discussion: Ontological and Epistemological Implications

The integration of TEK into ECEE allowed children to have a different view of nature. The Indigenous ontology and epistemology on which the two Elders drew from to teach environmental topics were different from the teacher's use of the curriculum. Under the teachings of the Elders, the com-munity had a strong relationship with nature, and they saw nature through this relationship (see Wilson, 2001, 2008). As Wilson (2001) noted, it is the relationship Indigenous people have with objects or ideas that is important. Similarly, Kasena's TEK was built through this relationship. Children were therefore exposed to an Indigenous worldview which may ultimately foster their relationship with nature. As France (1997, cited in Hart, 2010, p. 1) argued, "our worldviews affect our belief systems, decision making, assump-tions, and modes of problem-solving". The Elders' teaching provided a more informative understanding of the environment than the ECEE curriculum which focused on the cognitive development of Western science concepts.

From the Elders, ontologically, there was no clear separation between the people of Boania and nature. Similarly, Marshall et al. (2010) commented in the context of Canada that "From an Indigenous perspective, humans are inseparable from the rest of creation" (p. 174). This was also found in the community's relationship with the sacred trees, and other forms of creations. Harming an animal or a tree considered sacred means harming the whole village. There is a belief that harming sacred beings will bring curses on the entire village. By contextualizing teachings and highlighting how close the community's relationship to nature is, the Elder's teaching "reshapes abstract understandings of nature and land" (Seawright, 2014, p. 570).

In discussing the features of an Indigenous worldview, Simpson (2000) noted that the land is sacred and the relationship between people and the spiritual world is important. Hart (2010) also argued that "if the spiritual and sacred elements are surrendered, then there is little left of our philosophies that will make any sense" (p. 6). This was evident in the Elders' teachings. In teaching TEK, all forms of knowledge were taught holistically thereby avoiding what Nadadsy (1999) referred to as compartmentalization of TEK. Maurial (1999) stated, "one important basis of an Indigenous worldview expressed through Indigenous Knowledge is holisticity" (p. 63). Based on this, forms of knowledge are not divided into different disciplines. Therefore, the Elders employed religion and spirituality to explain environmental studies topics. Van der Walt (1997 cited in Thabede, 2008) argued that "African thought exists and differs from Western thought in that Western thought generally ignores the spiritual dimension of phenomena and focuses on the visible, measurable physical reality" (p. 235). As the Elders stated, no ceremony or activity can begin without pouring libation, which is the traditional method of seeking permission from the land. There was no separation between the spiritual and physical (see Datta, 2015). However, forms of knowledge were divided into different subjects in the ECEE curriculum. As a result, the teacher did not employ religion and spirituality in teaching environmental topics. Having been greatly influenced by DW-K, ECEE curriculum views religion and the environment as separate subjects. Unlike the teacher, the Elders employed religion in teaching about trees. Teachings such as these connect children more to nature.

Also, evident in the Elder's teaching of living and non-living things was that being able to grow, reproduce, move, and eat were not the only characteristics for classifying things as living and non-living. In other words, the Elders were more in tune with the Western science definition of living, as far as the tree was concerned, than the teacher who represented the Western science point of view to the children. The male Elder, for example, talked about the relationship that one has with things, by stating that a tree was more than just a tree. According to him, the relationship the people have with some sacred trees like the boabab and tamarind is sometimes equal to a family member. The act of viewing trees as sacred is not a new phenomenon and is definitely not limited to only the Kasena (Dafni, 2006, 2007). Nonetheless, teaching children to view trees as family members and to behave sustainably towards them is completely different from what the ECEE curriculum has on trees (see Ministry of Education, 2019). Teachings such as those from the Elders connect children more to nature. As Nxumalo and Villanueva (2019) observed, achieving sustainable behavior towards nature means changing the way EE is taught at the early childhood level. Where nature is often seen from an extractivist point of view (Nxumalo & Villanueva, 2019).

Again, the Elders emphasized the point that trees are caring and gave things (such as fruits, leaves, and herbs) freely to the community. For that, the community must in return care for the trees; this had great resonance with the children because some of the children had either eaten the fruit from or been treated with herbs from trees. This way of seeing nature is not present in the ECEE curriculum either (see Ministry of Education, 2019). Hence the children would not see trees as just trees after the teaching of TEK from the Elder. The Canadian anthropologist, Davis in a TED talk argued that our belief system determines how we relate to nature and our culture determines the environmental footprint we leave behind. According to Davis (2003) "a young kid from the Andes who is raised to believe that the mountain is an Apu spirit that will direct his or her destiny, will be a profoundly different human being and have a different relationship to that resource or that place than a young kid from Montana raised to believe that a mountain is a pile of rock ready to be mined" (10:10). Based on this analogy, by teaching children to view trees as family members in Boania, they may grow to have a different relationship with trees.

Also, I observed another important error in teaching science. The characteristics given for living things favor animals more than other forms of living things like trees. Perhaps if science was taught in the environment just like the teaching of TEK, the children and teacher might have noticed this error. The teaching of TEK was more engaging for the children and what they learnt was applicable to real-life situations. Taylor (1996) concluded that "learning occurs when students make connections between what is taught in the classroom and what happens in their daily lives outside of school" (p. 3). Furthermore, the integration of TEK into ECEE impacted the teacher's pedagogical practice. Having followed the children on some of the outdoor learning activities with the Elders, she started to employ the concept of nature giving things (medicine, fruits, shade, and food) freely to the community in her teachings. I observed that in teaching photosynthesis, the teacher emphasized how trees give oxygen freely to the community. In her final interview, she admitted that the Indigenous approach to teaching adopted by the Elders made concepts more relatable and easier for children to understand. Before this study, she did not think about nature that way. The teacher commented that Indigenous knowledges are sometimes such that they make lessons easy to grasp (teacher, interview [1] transcript, February 1st, 2020).

Lastly, the study showed children are capable of handling knowledge conflict. Dei (2000) concluded "to integrate Indigenous knowledges into Western academies is to recognize that different knowledges can coexist, that different knowledges can complement each other, and also that knowledges can be in conflict at the same time" (p. 120). The children showed that they were capable of managing conflict by adopting different strategies. Different hypotheses, the cultural Border Crossing Hypotheses by Aikenhead (1996), the Collateral Learning Hypothesis by Jegede (1995), the Contiguity Learning Hypothesis by Ogunniyi (1995), and the Cognitive Border Crossing Learning Model by Fakudze (2004) have been offered (in the context of science education) as to how children handle knowledge conflict when Indigenous knowledge is integrated into educational programs (see Fakudze & Rollnick, 2008). The children employed these different hypotheses to handle conflicting concepts and examples are given below.

I used the living and non-living things dichotomy to assess this by showing children pictures of a stone taught in class as a non-living thing and a stone on a shrine (with food, bloodstains, and chicken feathers on it) the children saw when the class visited the chief's palace. The children agreed with the teacher's teaching that a stone was a non-living thing. However, they also indicated that a stone on a shrine was a living thing because it was different from other stones. Pupil-9 (8-years) for example commented that stone can talk and eat because it is a shrine stone. It always speaks to the people (Pupil-9 interview transcript, March 23rd, 2020). Similarly, Pupil-2 responded that it was a shrine stone and it is different from other stones (Pupil-2 interview transcript, March 22nd, 2020).

These two responses represent the Collateral Learning Hypothesis (Jegede, 1995). A process whereby a student constructs, side by side, and with minimal interference and interaction, scientific and traditional meanings of concepts during and after a learning process (Jegede, 1995 as cited in Fakudze & Rollnick, 2008). By this, the students hold onto both belief systems that the stone is a non-living thing in the context of the classroom environmental studies and at the same time think a stone on a shrine is a living thing. I employed a second scenario on the use of the right and left hands. Among the Kasena of Boania, it is culturally disrespectful for a child to use the left hand to give or receive a gift from either a colleague or Elder. Hence, children are taught both in school and at home not to use the left hand to give or receive gifts. I asked the research participants what they would do if their teacher insisted (in the school environment) on them using the left hand instead of the right. Pupil- 9 again responded that he will use the left hand in school and the right hand at home (Pupil-9 interview transcript, March 23rd, 2020).

The response falls under the cultural border crossing hypothesis. This involves crossing borders from the subcultures associated with sociocultural environments into the subcultures of science (Aikenhead, 1996 as cited in Fakudze & Rollnick, 2008). Nonetheless, the Contiguity Learning Hypothesis by Ogunniyi (1995) is also applicable to this scenario where the two teachings (teacher's teaching in school and traditional teaching at home) compete, supplant, or dominate one another in/after the learning process, "depending on the worldview template serving as a frame of reference in a given context" (Fakudze & Rollnick, 2008, p. 88).

Indigenous Knowledges and Decolonizing Environmental Education

Kapyrka and Dockstator (2012) concluded, "Environmental education programs in general are taught from a Western perspective and typically do not engage with Indigenous knowledges" (p. 98). The question that emerges from the sections above is what does the centering of IKs, especially TEK in ECEE, mean for decolonizing EE which is the main theme of our book? In answering this question, it is important to note that decolonization as a process occurs differently under different contexts. Tuck and Yang (2012) stated that European colonialism occurred differently the world over. As a result, decolonization as a process will occur in different ways globally (Tuck & Yang, 2012). In this chapter, I argue that in the Ghanaian context, decolonizing EE can be achieved by incorporating IKs into the curriculum and pedagogy of EE. The issue of decolonizing EE centers on the question of "what is considered legitimate knowledge" (see Akena, 2012, p. 599). Akena observed, "European colonizers have defined legitimate knowledge as Western knowledge, essentially European colonizers' ways of knowing, often taken as objective and universal knowledge" (p. 600). Therefore, decolonizing EE means a paradigm shift, resisting, and the unsettling of Western science domination of the curriculum and pedagogy of EE.

Based on this, Ritchie (2012) observed that the inclusion of Indigenous perspectives in ECEE provides a valid counter-narrative to the DW techno-industrial emphasis that continues to damage our planet. Likewise, Dei and Simmons (2011) concluded that "Part and parcel of indigenous knowledge is resistance, survival; it is about operating counter-hegemonic to colonial Western forms of knowledge" (p. 109). The scholars further stated that due to the fact that decolonization is oriented as a counter-hegemonic process, "then we ought to speak about indigenous philosophies and the link to schooling and education" (p. 98). The inclusion of Indigenous worldviews in education develops awareness among students of the existence of IKs as legitimate knowledge (Dei, 2000). According to Dei, the creation of awareness among students of the existence of IKs is in itself a way of decolonizing education. It makes students aware that there are other legitimate ways of knowing. Hence decolonizing EE in Ghana means indigenizing or integrating Indigenous worldviews into EE. Dei and Simmon (2011) concluded "So in the African context, educational philosophies for decolonization must consider relevant knowledge that local peoples come to know as their own" (p. 99).

As shown by the study, integrating TEK into ECEE in Boania Primary School helped legitimize the knowledge. For example, children became aware that TEK (which we referred to as home knowledge) can also be taught in schools. It became evident that the dichotomy between living and non-living also includes relationships with nature and spirituality. Hence TEK challenged the DW-K in ECEE curriculum. "It provided a counter hegemonic discourse" (see Hoppers, 2002, p. ix). This was observed during one of the outdoor activities with the female Elder. One of the children, whose grandmother happened to be the Elder, said "I did not know that my grandmother was a teacher" (Pupil-4, participant observation notes, February 3rd, 2020). By inviting Elders into classrooms to teach ECEE, the children suddenly became aware that the DW definition of a teacher was actually questionable. This awareness is what decolonization aims to achieve. Respect for Indigenous knowledges will eventually be restored, once it is realized that they are equally important ways of knowing to Western scientific ways of knowing.

Besides, scholars (Bowers, 2001; Gruenewald, 2003) have argued that decolonization also seeks to recover and renew sustainable traditional and non-capitalist cultural behaviors. Based on this, Gruenewald (2003) defined decolonization as the process of "unlearning much of what dominant culture and schooling teach, and learning more socially just and ecologically sustainable ways of being in the world" (p. 9). By teaching children sustainable Kasena's traditional and non-capitalist IK, I believe that the study helped decolonize ECEE.

Conclusion

The need to develop children's love and respect towards nature has never been more urgent than now. As Pyle (2003) observed, people's sense of connection to nature "has paled, withered, and is finally failing" (p. 206). The objective behind this is that "early environmental experiences are vital in developing a relationship with the environment and developing environmental concern" (Barratt Hacking, Barratt, & Scott, 2007 cited in Boileau, 2013, pp. 142–143). Hence, as Pyle (2003) noted "there is no longer any doubt that a strong individual sense of connection to nature and natural processes is utterly essential to the healthy coexistence of humans with their biological neighbours and physical setting" (p. 206).

Ritchie (2012) argued that TEKs are more suited to connect children to nature because they emphasize "respectful interdependence with nature" (p. 63). Besides when employed in ECEE, the teaching of TEK is less structured and gives children the freedom to explore nature at their own pace. This child-led approach to learning supports fit with the goals of ECEE. As the NAAEE (2010) noted "The approach to environmental education for early childhood learners is less about organization of graduated achievements and more about free discovery on each child's own terms" (p. 3).

Importantly, TEK helps in decolonizing ECEE by challenging DW-K as the only legitimate environmental knowledge. Therefore, this chapter broadly concludes that relying on DW science alone to deliver EE is not the best approach to achieving sustainability and EE must find ways to incorporate IKs and cultures. The same conclusion has been arrived at by other scholars (see Agyeman, 2002;; Korteweg & Russell, 2012; Marouli, 2002; Martusewicz et al., 2011; Root, 2010; Sauvé, 1997; Taylor, 1996; Wilson, 2001). In this chapter, I give further evidence of why and how TEK can be integrated into ECEE. An important aspect is the use of Elders to teach, with the teacher, curricular concepts. A second important aspect is that the Elders provoked the children's curiosity and learning by taking them into the outdoors, to the places where the concepts are best learned.

References

Agyeman, J. (2002). Culturing environmental education: From First Nation to frustration. *Canadian Journal of Environmental Education, 7*(1), 5–12.

Aikenhead, G. (1996). Science education: Border crossing into the subculture of science. *Studies in Science Education, 26,* 1–52.

Akena, F. A. (2012). Critical analysis of the production of Western knowledge and its implications for Indigenous knowledge and decolonization. *Journal of Black Studies, 43*(6), 599–619. <https://doi.org/10.1177/0021934712440448>.

Barratt Hacking, E., Barratt, R., & Scott, W. (2007). Engaging children: Research issues around participation and environmental learning. *Environmental Education Research, 13*(4), 529–544.

Bartlett, C., Marshall, M., & Marshall, A. (2012). Two-eyed seeing and other lessons learned within a co-learning journey of bringing together Indigenous and mainstream knowledges and ways of knowing. *Journal of Environmental Studies and Sciences, 2*(4). <https://doi.org/10.1007/s13412-012-0086-8>.

Boileau, E. Y. S. (2013). Young voices: The challenges and opportunities that arise in early childhood environmental education research. *Canadian Journal of Environmental Education, 18,* 142–154.

Born, P. (2018). Regarding animals: A perspective on the importance of animals in early childhood environmental education. *International Journal of Early Childhood Environmental Education, 5*(2), 46–57.

Bowers, C. A. (2001). *Educating for Ecojustice and Community.* Athens: The University of Georgia Press.

Dafni, A. (2006). On the typology and the worship status of sacred trees with a special reference to the Middle East. *Journal of Ethnobiology and Ethnomedicine, 2*(26), 1–14.

Dafni, A. (2007). Rituals, ceremonies, and customs related to sacred trees with a special reference to the Middle East. *Journal of Ethnobiology and Ethnomedicine, 3*(9), 1–15.

Datta, R. (2015). A relational theoretical framework and meanings of land, nature, and sustainability for research with Indigenous communities. Local Environment: *The International Journal of Justice and Sustainability, 20*(1), 102–113.

Davis, W. (2003, February). Dreams from endangered cultures [video file]. Retrieved from: <https://www.ted.com/talks/wade_davis_dreams_from_endangered_cultures#t-635080>. September 24, 2020.

Dei, G. J. S. (2000). Rethinking the role of Indigenous knowledges in the academy. *International Journal of Inclusive Education, 4*(2), 111–132. <https://doi.org/10.1080/136031100284849>.

Dei, G. J. S., & Simmons, M. (2011). Indigenous knowledge and the challenge for rethinking conventional educational philosophy: A Ghanaian case study. *Counterpoints, Regenerating the Philosophy of Education: What Happened to Soul?, 352,* 97–111.

Delvaux, C. Sinsin, B., Darchambeau, F., & Van Damme, P. (2009). Recovery from bark harvesting of 12 medicinal tree species in Benin, West Africa. *Journal of Applied*

Ecology, *46*, 703–712. <https://doi.org/10.1111/j.1365-2664.2009.01639.x>.Elvs tam, A. & Fleischer, S. (2018). Ghanaian teacher students' view on using outdoor pedagogy when teaching natural science (Degree Project). Retrieved from <https://pdfs.semanticscholar.org/74ce/fe7338e729a9acc80bc889c91bdbfe2debe7.pdf>. August 3, 2020.

Fakudze, C. G. (2004). Learning of science concepts within a traditional socio-cultural environment. *South African Journal of Education*, *24*(4), 270–277.

Fakudze, C., & Rollnick, M. (2008). Language, culture, ontological assumptions, epistemological beliefs, and knowledge about nature and naturally occurring events: Southern African perspective. L1– *Educational Studies in Language and Literature*, *8*(1), 69–94.

France, H. (1997). First Nations: Helping and learning in the Aboriginal community. *Guidance and Counseling*, *12*(2), 3–8.

Global Environmental Education Partnership [GEEP] (2019). Ghana. Retrieved from <https://thegeep.org/learn/countries/ghana>. October 12, 2019.

Greenwood, M., de Leeuw, S., Lindsay, N. M., & Reading, C. (2015). *Determinants of Indigenous peoples' health in Canada: Beyond the social*. Toronto: Canadian Scholars Press.Gruenewald, D.A. (2003). The best of both worlds: A critical pedagogy of place. *Educational Researcher*, *32*(4), 3–12.

Hart, M. A. (2010). Indigenous worldviews, knowledge, and research: The development of an indigenous research paradigm. *Journal of Indigenous Voices in Social Work*, *1*, 1–16.

Hoppers, C. A. O. (Ed.). (2002). *Indigenous knowledge and the integration of knowledge systems:Towards a philosophy of articulation*. South Africa: New African Books (Pty) Limited.

Iwama, M., Marshall, A., Marshall, M., & Bartlett, C. (2009). Two-eyed seeing and the language of healing in community-based research. *Canadian Journal of Native Education*, *32*, 3–23.

Jegede, O. J. (1995). Collateral learning and the eco-cultural paradigm in science and mathematics education in Africa. *Studies in Science Education*, *25*, 97–137.

Kapyrka, J., & Dockstator, M. (2012). Indigenous knowledges and western knowledges in environmental education: Acknowledging the tensions for the benefits of a "two-worlds" approach. *Canadian Journal of Environmental Education*, *17*, 97–112.

Kim, E. A., & Dionne, L. (2014). Traditional ecological knowledge in science education and its integration in Grades 7 and 8 Canadian science curriculum documents. *Canadian Journal of Science, Mathematics and Technology Education*, *14*(4), 311–329. <https://doi.org/10.1080/14926156.2014.970906>.

Kim, E. J. A., Asghar, A., & Jordan, S. (2017). A critical review of traditional ecological knowledge in science education. *Canadian Journal of Science, Mathematics and Technology Education*, *17*(4), 258–270. <https://doi.org/10.1080/14926 156.2017.1380866>.

Kimmerer, R. W. (2002). Weaving traditional ecological knowledge into biological education: A Call to action. *BioScience, 52*(5), 432–438.

Kimmerer, R. W. (2012). Searching for synergy: Integrating traditional and scientific ecological knowledge in environmental science education. *Journal of Environmental Studies and Science, 2*, 317–323. <https://doi.org/10.1007/s13412-012-0091-y>.

Korteweg, L., & Russell, C. (2012). Decolonizing + Indigenizing = moving environmental education towards reconciliation. *Canadian Journal of Environmental Education, 17*, 5–14.

Marsh, T. N., Cote-Meek, S., Toulouse, P., Najavits, L. M., & Young, N. L. (2015). The application of two-eyed seeing decolonizing methodology in qualitative and quantitative research for the treatment of intergenerational trauma and substance use disorders. *International Journal of Qualitative Methods, 14*, 1–13. <https://doi.org/10.1177/1609406915618046>.

Marshall, A., Marshall, M., & Iwama, M. (2010). Approaching Mi'kmaq teachings on the connectiveness of humans and nature. In S. Bondrup-Nielsen, K. Beazley, G. Bissix, D. Colville, S. Flemming, T. Herman, M. McPherson, S. Mockford, & O'Grady (Eds.), *Ecosystem Based Management: Beyond Boundaries. Proceedings of the Sixth International Conference of Science and the Management of Protected Areas*, 21–26 May 2007 (pp. 174–177), Acadia University, Wolfville, Nova Scotia. Science and Management of Protected Areas Association, Wolfville, NS. Retrieved from: <http://integrativescience.ca/uploads/articles/2010-Beyond-Boundaries-ecosystem-based-management-Marshall-Iwama-SAMPAA-2007-proceedings.pdf>. July 26, 2020.

Marouli, C. (2002). Multicultural environmental education: Theory and practice. *Canadian Journal of Environmental Education, 7*(1), 26–42.

Martusewicz, R., Edmundson, J., & Lupinacci, J. (2011). *Eco justice education toward democratic and sustainable communities.* New York: Routledge.

Maurial, M. (1999). Indigenous knowledge and schooling: A continuum between conflict and dialogue. In L. M. Semali & J. L. Kincheloe (Eds.), *What is Indigenous knowledge: Voices from the academy* (pp. 59–77). New York: Falmer Press.

McCarter, J., & Gavin, M. C. (2011). Perceptions of the value of traditional ecological knowledge to formal school curricula: opportunities and challenges from Malekula Island, Vanuatu. *Journal of Ethnobiology and Ethnomedicine, 7*(38). Retrieved from: <http://www.ethnobiomed.com/content/7/1/38>.

Ministry of Education. (2019). Kindergarten curriculum (KG 1&2). Kindergarten curriculum for preschools. National Council for Curriculum and Assessment. Ministry of Education, Accra, Ghana.

Nadasdy, P. (1999). The politics of tek: Power and the "integration" of knowledge *Arctic Anthropology, 36*(1/2), 1–18.

Nelson, N., Pacini-Ketchabaw, V., & Nxumalo, F. (2018). Rethinking nature-based approaches in early childhood: Common worlding practices. *Journal of Childhood Studies, 43*(1), 4–14. <https://doi.org/10.18357/jcs. v43i1.18261>.

North American Association for Environmental Education (NAAEE). (2010). *Early Childhood Environmental Education Programs: Guidelines for Excellence.* Washington, DC: Author.

Nxumalo, F. (2019). *Decolonizing place in early childhood education.* New York: Routledge.

Nxumalo, F., & Cedillo, S. (2017). Decolonizing 'place' in early childhood studies: Thinking with Indigenous onto-epistemologies and Black feminist geographies. *Global Studies of Childhood, 7*(2), 99–112.

Nxumalo, F., & Villanueva, M. (2019). Decolonial water stories: Affective pedagogies with young children. *The International Journal of Early Childhood Environmental Education, 7*(1), 40.

Ogunniyi, M. B. (1995). World view hypothesis and research in science education. *Proceedings of the annual meeting of the Southern African Association for Research in Mathematics and Science Education* (pp. 613–624).

Parrota, T. A., & Trosper, R. L. (Eds.). (2012). *Traditional forest related knowledge: Sustaining communities, ecosystems and biocultural diversity.* New York: Springer.

Patrut, A., Von Reden, K. F., Lowy, D. A., Alberts, A. H., Pohlman, J. W., Wittmann, R., Gerlach, D., Xu, L., & Mitchell, C. S. (2007). Radiocarbon dating of a very large African baobab. *Tree Physiology, 27,* 1569–1574.

Pence, A., & Shafer, J. (2006). Indigenous knowledge and early childhood development in Africa: The early childhood development virtual University. *Journal for Education in International Development, 2*(3), 1–16.

Pyle, R. M. (2003). Nature matrix: Reconnecting people and nature. *Oryx, 37,* 206–214.

Reid, A., Teamey, K., & Dillon, J. (2002). Traditional ecological knowledge for learning with sustainability in mind. The Trumpeter. *Journal of Ecosophy, 18*(1), 113–136.

Ritchie, J. (2012). Titiro Whakamuri, Hoki Whakamua:1 Respectful integration of Maori perspectives within early childhood environmental education. *Canadian Journal of Environmental Education, 17,* 62–79.

Root, E. (2010). This land is our land? This land is your land: The decolonizing journeys of white outdoor environmental educators. *Canadian Journal of Environmental Education, 15,* 103.

Sauvé, L. (1997). Canada – environmental education and the sustainable-development perspective. Available at: <http://allies.alliance21.org/polis/spip.php?article120>. Last accessed June 16, 2017.

Seawright, G. (2014). Settler traditions of place: Making explicit the epistemological legacy of white supremacy and settler colonialism for place-based education. *Educational Studies, 50*(6), 554–572.

Simpson, L. (2000). Anishinaabe ways of knowing. In J. Oakes, R. Riew, S. Koolage, L. Simpson, & N. Schuster (Eds.), *Aboriginal health, identity and resources* (pp. 165–185). Winnipeg, Manitoba, Canada: Native Studies Press.

Swart, E. R. (1963). Age of the baobab tree. *Nature, 198,* 708–709.

Taylor, D. E. (1996). Making multicultural environmental education a reality. *Race, Poverty & the Environment, 6*(2/3), 3–6.

Thabede, D. (2008). The African worldview as the basis of practice in the helping professions. *Social Work/Maatskaplike Werk, 44*(3), 233–245.

Tilbury, D. (1994). The critical learning years for environmental education. In R. A. Wilson (Ed.), *Environmental education at the early childhood level* (pp. 11–13). Washington, DC: North American Association for Environmental Education.

Tuck, E., & Yang, K. W. (2012). Decolonization: Indigeneity. *Education & Society, 1*(1), 1–40.

Van der Walt, B. J. (1997). *Afrocentric or Eurocentric? Our task in a multicultural South Africa.* Potchefstroom: Potchefstroom University.

Wilson, R. (1994). *Environmental education at the early childhood level.* Washington, DC: North American Association for Environmental Education.

Wilson, S. (2001). What is an indigenous research methodology? *Canadian Journal of Native Education, 25*(2), 166–174.

Wilson, S. (2008). *Research is ceremony: Indigenous research methods.* Black Point, NS, Canada: Fernwood Publishing.

Wood, P. J. (1992). The botany and distribution of Faidherbia albida. In R. J. Vandenbelt (Ed.) *Faidherbia albida in the West African semi-arid tropics* (pp. 9–17). Workshop.

"Earth Sorrow" (2021)
Elsa McKenzie

7. Naturalized Places, Indigenous Epistemology, and Learning to Value

Janet McVittie & Marcelo Gules Borges

We live, love, work, and play here in misâskwatômina, on Treaty 6 territory, home to the temperate grasslands, one of the world's most threatened ecosystems. We honour the people who lived sustainably on this land from time immemorial, the neyonawak, saulteaux, dakota, lakota, nakota peoples and homeland of the Métis peoples.

Introduction

The purpose of this chapter is to share the learning that has taken place within a naturalized area in Treaty 6 territory in the land now known as Canada. The naturalized area, the Prairie Habitat Garden (PHG) next to the College of Education, University of Saskatchewan, has become a place for teaching pre-service teachers, supporting in-service teachers, and teaching children from the surrounding city and rural areas, about Indigenous peoples, the ecology of the local prairies, Indigenous perspectives and relationships with the land, and about other entities that are intricately relationally tied to the land. The ontology and epistemology of Indigenous peoples of this territory focus on relationships, rather than on entities. Both exist—entities and relationships. However, where entities are foregrounded within the Dominant Western Worldview (DWW), relationships are foregrounded for this group of Indigenous peoples' ontology.

We believe that there is sufficient information on what we must do to interrupt climate change and species loss. Western scientists have clearly identified the problems (Intergovernmental Panel on Climate Change, 2018; World Wildlife Federation, 2018), and a number of solutions have been proposed, most of them technological. A significantly different solution would be to take up a relational ontology, where every entity is seen as a necessary part of the whole. We two authors believe that what is needed is a shift in how the world is viewed. We two authors believe that student and teacher values will change as they come to understand the multiple relationships amongst various entities, as they come to see the world more as relational, shifting the focus from the nexus points that are the entities, to the connecting threads which weave the fabric of understanding and being. Seeing the world as a fabric of life (Deloria, 2001) is a challenging shift from DWW ways of knowing. In this chapter, the authors present how the PHG has shifted the authors' perspectives on valuing, but also the perspectives of pre-service teachers. The researchers and the pre-service teachers who worked in the garden have come to value how Indigenous peoples know/knew the land, live/lived with the land, value/valued the land. For many instructors, teaching about Indigenous people involves facts, dates, injustices committed. It is important to acknowledge past wrongs. However, in this chapter, we are searching for a way to bring about an ontological shift for the pre-service teachers we work with, such that they live in relational ways with the land, and the original peoples of the land, and thereby learn with the land. This learning with the land, and focus on relationships has made us and the pre-service teachers we work with more aware of the need to value Other[1]; indeed, we have learned to see Other as part of us: we are all related (Cajete, 2005).

Perspectives

Settler Ancestry and Relational Ontology

We co-authors are in an awkward situation. We are of settler ancestry in our countries (Marcelo from Brazil, Janet from Canada, with the research conducted in Canada), living on land stolen from Indigenous peoples. Neither of us wishes to appropriate Indigenous knowledge, or ways of knowing; further, we recognize we cannot speak on behalf of Indigenous peoples. What we have done to develop greater understanding of Indigenous ontology is (1) drawn on Indigenous scholars from the great plains of North America, to develop understanding of this group of Indigenous peoples' perspectives; (2) worked with Indigenous Elders for building replicas of Indigenous artefacts in the PHG, and to learn more about relational ontologies as lived; (3) acknowledge that we are of settler ancestry and want to understand more, as we live in humility with our hosts who have shared this land with us; and (4) recognize that there are multiple Indigenous ontologies—the peoples from the Great Plains of North America have different ontologies than the people of the mountain chaparrals of Brazil, and every individual has a unique ontology and epistemology. McGregor (2005) noted that there is a contrast between Indigenous academics who write about traditional ecological knowledge (TEK), and those who actually live it. She argued:

> Minobimaatisiiwin is so much more than knowledge about how to live sustainably. Rather, it is living sustainably. It is not just about understanding the relationship with Mother Earth, it is the relationship itself. Academics are not incorrect to say that Indigenous people all over the world possess knowledge that is sustainable in nature and can be helpful to broader society. Indigenous people have been saying this for years (Clarkson et al., 1992). TEK includes specific knowledge that can be described as ecological or environmental, but it is much more than that. (p. 104)

Her argument was against the co-opting of traditional knowledge by Western scientists, that the knowledge thereby would be de-contextualized, taken away from the knowledge holders, and generalized for other contexts in which it might not apply.

We two authors have been pursuing understanding of this complex and contextual concept of relational ontology. We concur with McGregor's argument that humans must live with the land, to truly understand how to live sustainably. We believe that living sustainably means to adopt a relational ontology. If Indigenous people living on the land do not treat their ecosystems well, they will starve. Adding a grocery store to a community means the community can be separated from the consequences of their immediate poor environmental practices. As well, spiritual and ethical connections amongst all entities do not develop unless people live alongside their food; hasty purchases in grocery stores do not support these connections. Thus, we two urban-living researchers cannot claim the depth of knowledge that McGregor (2005) speaks of.

We wonder also about how deep and how inclusive relational ontologies are amongst Indigenous peoples of the Great Plains of North America. There are many perspectives; not all Indigenous individuals are oriented toward the land. In two recent court cases in Canada, and one in Australia, there was consultation with Indigenous peoples (albeit, shallow consultation), and different perspectives emerged from amongst the people. For example, in the TransMountain Pipeline dispute in Canada, Chief Tony Alexis and Paul Poscente (2019) put forward a case for Indigenous peoples to own the pipeline, noting:

> For Canada, this is an opportunity to more positively redefine the relationship with Indigenous communities as it relates to major resource development. For Indigenous communities, asset ownership is a significant step along the spectrum from managing poverty to managing wealth. While TMX is working

toward getting shovels in the ground for this multibillion-dollar project, we are working with our communities in Alberta and B.C. to form a pragmatic, measured approach in order to make a credible offer when the pipeline investment becomes feasible. (Alexis & Poscente, 2019)

In Australia, also, there is controversy amongst the Indigenous peoples regarding extractive and anti-environmental development of land. Adani Corporation has just been granted an Indigenous land use agreement (ILUA), because, out of twelve Indigenous groups who have claims to the land, seven have agreed to the extinguishing of their rights to manage the land (Doherty, 2019).

In both Canada and Australia, as seen from these examples, settlers have been "developing" the land, extracting resources, and becoming wealthy, for centuries, while Indigenous peoples have been pushed to the side, systematically given the poorest land, had their children stolen and murdered, and have lived in poverty. The ability to participate in the economy is attractive, especially with some examples in Canada such as the Onion Lake First Nation, which owns and develops its own oil reserves (<https://onionlake.ca/>). Onion Lake has become a wealthy First Nation, able to provide employment for many of its people, as well as building health care and education services for its people. Thus, from the environmental perspective, taking up an Indigenous relational ontology must be seen for what it is: valuing the way that Indigenous people valued (in the past, and many currently) the land was what kept ecosystems healthy; environmentalists side with the environment. However, are environmentalists co-opting Indigenous peoples to take up their cause? As Canada attempts to reconcile for the horrors visited upon Indigenous peoples, the country will have to ensure Indigenous peoples have access to the land, including the resources, and for settlers to get out of the way of Indigenous peoples as they—the Indigenous peoples—decide how best to move forward with these resources. This is hard for us two researchers, in our environmental bones. Nonetheless, settlers have long been wealthy, and Indigenous peoples have suffered horribly. And Indigenous people cannot manage the environment any worse than settlers have.

This leads to another tension for teachers of settler ancestry: can we teach from an Indigenous perspective? The answer is a simple "No". The very definition of perspective is that it is from a particular person's viewpoint. A non-Indigenous person cannot teach from an Indigenous perspective. Indigenous perspectives vary widely. We, as researchers of settler ancestry, can teach *about* multiple Indigenous perspectives, but we have to acknowledge that different individuals and different groups of individuals will have unique beliefs about the land. Only an Indigenous person can have an Indigenous perspective, and that perspective will always be more or less unique to that one person, as able to be imagined within the cultural hegemony. Thus, although each individual will imagine the world in a specific way, there are commonalities, and there are things that cannot be imagined. For example, not long ago, Indigenous people could not imagine the kind of damage that has been done to the land (and to the people Indigenous to that land) in the name of wealth creation.

Sovereignty

English, the dominant language in this place, is noun based, and thus places greater value on entities. Entities do exist; they are valuable. However, the valuing of individual entities over all else has led to self-centered destruction of other peoples, and of the environment. This valuing of entities derives from or supports DWW ontologies and epistemologies, as pointed out by Seawright (2014). Seawright noted the few steps it took for John Lock to saying only "developed" land had value, to justifying ownership of other humans. Entities matter, but without noticing the connections among them, some entities matter a lot more than others. This leads to the tension regarding beliefs about land ownership. From our readings of concepts of land "ownership", we believe that Indigenous peoples in North America had different ideas than those brought to North America from Europe. Two concepts existed in Europe: public land (shared among the people), and private property (belonging to one individual or one family unit). The concept of "public" land, of the commons, was usually dismissed by settlers, since many of them had had their public land stolen from them while in Europe (Patel, 2009). The newcomers to North America tended to seek private ownership of land. However, neither of these concepts, neither the commons nor private ownership, were much used in North America, prior to colonization. Here, the people belonged to the land (Calderon, 2014; Seawright, 2014). Therefore, for Indigenous peoples in North America, land was neither publicly owned, nor privately. This, however, does not equate to *terra nullius*, the concept that land was not occupied nor managed, and could therefore be claimed by European settlers.

For Indigenous peoples on the great plains of North America, all entities existed only in relationships, and humans lived in humility with all other entities. These entities all belonged to the land. As Annie Peaches said: "The land is always stalking people. The land makes people live right. The land looks after us. The land looks after people" (cited in Basso, 1995, p. 38). Land was not able to be bought, sold, claimed, stolen; rather all beings were intricately connected through relationships beginning with the land (Seawright, 2014).

Cajete (2005), in the following explanation, articulated the notion of relationality as being at the heart of Indigenous teaching and learning: "*Mitakuye Oyasin* (we are all related) is a Lakota phrase that captures an essence of tribal education because it reflects the understanding that our lives are truly and profoundly connected to other people and the physical world" (p. 70). Indigenous approaches to education center on the respect and care for the land *in situ* (Bowers, 2001, 2010; Calderon, 2014; Coulthard, 2010; Deloria, 2001; Ritchie, 2015; Simpson, 2011, 2014; Wildcat, 2001, Wildcat et al., 2014). As Deloria (2001) noted: "The best description of Indian[2] metaphysics was the realization that the world, and all its possible experience, contributed a social reality, a fabric of life in which everything had the possibility of intimate knowing relationships because, ultimately, everything was related" (p. 2).

This particular tension addresses sovereignty. Sovereignty means to have rights and power over a people and/or a place. Yet Indigenous people claimed that the land actually "owned" them, or, at the most, "ownership" was reciprocal; therefore, was there any sovereignty prior to the arrival of settlers? Although the pope and European nobility would have claimed the land for themselves regardless, the concept of *terra nullius* was invoked: the land was not owned by Indigenous peoples, and therefore, was available to be claimed by European settlers. *Terra nullius* was justified along a number of lines, including the lack of sovereigns (kings), a perceived lack of "development", and that any group of persons who were not Christian were not human, and therefore, the land was open to be claimed by Europeans.

Although in the past, Indigenous people of the Great Plains of North America might have felt they belonged to the land, rather than the other way, this must not be used as any part of any current argument to extinguish Indigenous rights to the land. Indigenous peoples had power with the land, in the sense that they managed the land in consideration of all entities there. This attitude toward responsible "sovereignty" is demonstrated in research indicating that when management of land is given to Indigenous peoples, the land becomes or remains ecologically healthy (Renwick et al., 2017; Schuster et al., 2019). Schultz et al. (2019) argued: "in remote regions of Australia, Indigenous Land Management (ILM) is being recognised for its potential to restore ecosystems, revitalize languages and enhance health and well-being for Indigenous people" (p. 171). In other words, Schultz et al. argued that having Indigenous people manage the land worked both ways—benefiting the environment, and the people; their study supported their hypothesis, thus demonstrating that greater understanding of and control of the land could support reconciliation.

The Research

The Question

What we can see from Indigenous academics who have written on their ontology is that, for the Indigenous peoples of the Great Plains of North America, their ontology was (and for many still is) based on relationality, rather than on entities. Within this way of knowing, they were able to live sustainably with the land and the other entities therein. The question this chapter addresses is: *in what ways might pre-service teachers of settler ancestry come to value land, relationality, and the Indigenous peoples who developed and currently live these ontologies?* We wondered if these pre-service teachers could come to understand/live in a different way with land through living in relationship with a particular piece of land? Does this change their way of relating to land in general? It is not Indigenous people we wish to change; it is those who have been well schooled in DWW. But we wanted to develop theory around whether pre-service teachers of settler ancestry could take up an Indigenous relational ontology, while ensuring they honor that this ontology came from the local Indigenous peoples.

Methodology

Using constructivist grounded theory methodology (Bryant & Charmaz, 2010; Kenny & Fourie, 2014, 2015; Seldén, 2005), we interviewed pre-service teachers who worked as summer students in the garden (from 2016 to 2018). Grounded theory draws on the data to support the development and/or modification of theory. Along with Bryant and Charmaz (2010), we acknowledge that the researcher is inevitably value laden and biased, which interrupts the potential for a completely open interpretation towards theory building. Following the recommendations of Bryant and Charmaz, we took care to develop awareness of our biases. We were searching for a theory to determine in what ways pre-service teachers of settler ancestry might come to value land, relationality, and the Indigenous peoples who developed and live these ontologies.

Five pre-service teachers had been employed to work in the garden for a sustained time of 16 weeks, weeding out invasive species, planting more native species, creating and carrying out or observing workshops with K to 8 (ages 5–14 years) students, reading books and papers about native prairies and Indigenous peoples, and occasionally working with Indigenous Elders. Three of the five summer students were interviewed, and constructivist grounded theory methods were used to search for their learning regarding Indigenous perspectives; as well, the summer students' blogs were examined in an attempt to determine their learning.

The two authors of this chapter then discussed and critiqued one another's interpretations of data, working to keep our minds open to potentialities, as well as looking for points of difference amongst the data. We sought in what ways we, all of us well schooled in DWW, saw and lived that piece of the land, the entities, and the relationships, as we struggled to understand the land from a relational and Indigenous perspective. Through this process of organizing, analyzing, and discussing the data, a text emerged, illustrating the themes the students held. We hoped that the theory we sought regarding involvement with a naturalized fragment of native prairie would support pre-service teachers to understand an Indigenous ontology of relationality, and that they would then carry that into their teaching practice.

Data Sources

The roles the pre-service teachers took up in the PHG included maintaining the garden by removing invasive weeds, and planting new native plants; as well, they each had the opportunity to work with at least one Elder, who taught a group of elementary school children at an event. The pre-service teachers were in the garden daily, often up to six hours a day, five days a week, for a total of sixteen weeks, during Saskatchewan's summer. When not in the PHG, they were reading papers and books by Indigenous authors about the land. The sustained time in the PHG led to them noticing new species of plants and animals they had not known about, and these were commented on in their blogs, often with photographs attached. The pre-service teachers provided the data, which included interviews with three of the pre-service teachers, all of whom had graduated and were teaching; supplemental to these are blogs which several students kept during their summer work. The interview questions addressed their backgrounds, their experiences with the garden through a course on teaching in one's own place, their experiences in the garden as summer students (with probes addressing Indigenous features and perspectives), the role of humans in the world, and the value of the PHG for pedagogy. We wondered how these pre-service teachers took up Indigenous concepts of valuing and of relationality.

Results

The Site

The PHG was created in 2005; since then, the first author and a colleague obtained a substantial grant, allowing for major renovations, clearing out the invasive plants, planting a greater variety of Indigenous plants, and incorporating, with the support of Indigenous elders, various replicas of Indigenous artefacts (https://www.youtube.com/watch?v=tO_yVoZVCbE). The first replica built was of an Earth turtle (2014), then a bison head (2015), then an effigy of the sun (2016), a chicken lodge (2017), and a Celestial Circle (2017) were added. As well, a set of logs have been made into a circular seating area, and a circular picnic table with an interpretive Medicine Wheel on it have been added.

The Earth turtle represents a story told by many North American Indigenous peoples about how Sky Woman (the first human) fell to earth. In the version of the story by Kimmerer (2013), the emphasis is on the cooperation amongst all the entities in saving Sky Woman and creating the land.

Another replica within the garden is the Celestial Circle, a circle of stones, with arms indicating (for the one in the garden) sunrise and sunset at both solstices, as well as an arm pointing to the next nearest Celestial Circle. The oldest of the actual extant celestial circles that have been found is approximately 7,000 years old, and there are Celestial Circles which track major stars as well as the sun (King, 2017).

The grant allowed for additional environmental features to the garden as well. For example, a swale was enhanced in a low lying area of the garden, and where the swale drops deeply, a bridge made of re-purposed lumber was built over it. The swale is an opportunity to teach about water, rain, and the land; when "developers" make the ground impermeable with concrete or tarmac, rain must flow into rivers (washing pollutants from the streets with it), increasing flooding. When the ground remains permeable, and there are low lying areas (swales) to absorb the rainwater, the land is healthier, and floods are less likely.

The Data

The rich data from the pre-service teachers' blogs and the interviews meant we could generate theory regarding sustained involvement with the land. Two of the pre-service teachers wrote about their experiences in blogs. As they interacted daily within the PHG, they developed connections with the different species there. For example, one student revealed in her blog about crab spiders which camouflage themselves by taking on the color of the flower they are in so as to capture prey (the pre-service teacher who found the crab spider observed it daily, noticing how its color intensified to completely disguise itself in its yellow cone-flower home); another student noticed indigenous bees which hive in the ground, near where the fertilized queen has hibernated solo for the winter. The summer students came to see the visitors to and the inhabitants of the garden as nexus points in webs of relationships.

One summer student developed a relationship with a warbler family. She noticed the nest with four eggs as she was removing invasive (but native) wild roses. The warbler nest was on one of the stems—the thorns and the density of the growth of the roses provided shelter for the nest. She quickly re-inserted the rose stem into the ground, and left that area alone; however, she had removed most of the surrounding roses already. She noticed the parent warblers, flying in and out of the rose bushes in that area, and so she crept back in to check on the eggs a few days later. Four naked baby birds, with huge beaks! Whenever she snuck in to see them, they would lift up their heads and open their beaks wide, as if they thought she would feed them. She limited her visits to twice a day, so the parents could go in and out of the wild rose bushes, undisturbed by her. Then one day over the noon hour, the baby warblers disappeared. She realized the resident crow was likely responsible. However, the crows are important in the garden as well. This pre-service teacher learned the complexity of the relationships: the density of and thorns on the wild roses provided a safe home for the warblers. By removing the rose bushes from this part of the garden, we had put the young warblers at risk.

Through examination of the interviews, three themes were identified: learning to value self; learning to value relationships; learning to value and teach Indigenous relational ontology. The summer students, now having completed their degrees and teaching kindergarten to grade twelve students, revealed the potency of the learning experiences in the PHG. The idea of relationality emerged in a variety of narratives, with each of the three interviewees unveiling maturity in terms of their ability to translate the idea of relationality into teaching experiences.

The Themes

Learning to Value Time Alone in the Garden for Self. All three pre-service teachers emphasized learning to value time alone in the garden for self. The time away from social media, away from other humans, as they worked with the physical materiality of pulling out invasive grasses or planting new plants, allowed them to think. This time to think contributed to their own developing identities, as well as to their health. One commented how she came to understand her own history as a refugee and immigrant to Canada, noticing for the first time that it was similar in some ways to the history of Indigenous people. She said:

> It was quite transformational. I would say, there was ... a lot of self-reflection and self-discovery that I was going through. Just, my past has not been the easiest, coming from an immigrant family, very low income, and had a very hard

time establishing ourselves in Canada. And then there was just a lot through my journey that I had experienced and coming into the garden was just a place that kind of helped me make sense of my own world. I loved spending time in there, and working with my hands, and just kind of letting my mind wander off.

In Canada, she had felt at first that she did not belong, that she was a (somewhat) unwelcome guest, and she recognized while working in the garden, for herself a surprise, that Indigenous people had been marginalized in their own country, pushed to the periphery socially, as well as materially. She had spent her Canadian childhood interacting with children of settler ancestry and recent immigrants, for whom racism against Indigenous peoples was the norm, and so connecting to the history of Indigenous people was a transformation.

As well, the material physicality of having her hands in dirt contributed to letting her mind wander: material reality and time to reflect were integral to her transformation. The PHG experience, for her, was self-healing, as she connected to her past and present, to the past and present of this land and the people indigenous to here.

Participant 2 noted:

I thought it was really relaxing and I enjoyed coming and having that time to myself. And just being free to think while I am working. And I got to see the animals that would come by, like Richardson ground squirrels, bats, snakes, mice, and geese. I really enjoyed it. I thought it was really a great experience.

As with Participant 1, this pre-service teacher found the physical materiality of the garden important, and the time in the garden allowed her to let go of other stressors that normally interrupted her ability to learn and grow. She commented as well on her relationship with the animals in the garden. Participant 2 noted that, as a child, she recognized that anything she did had repercussions into the wider world. The garden gave her the time to reflect, to consider her identity, to think about what she valued in the present moment, free of financial worries for a time.

Participant 3 actually measured, in a Western science approach, the effect of the garden on his health:

I don't know if you remember, but I had my fit-bit on all the time and my resting heart rate started to drop when I was working in the garden. The mental well-being, I was happier all the time there, a great sense of fulfillment with the work. I was happy to be involved with the project. The social aspect, felt like a part of the community, recognizing the importance of community and having a role in the community. Also very spiritual, not an idea of God so much, but I developed a greater connection to - not the creator really, but to creation. In relation to my settler identity. Is this tokenism? A strong sense of connection to creation.

The mere physical act of walking into the PHG supported his physical well-being. His comment about his connection to "creation" was pertinent, as was his worry about being tokenistic in talking about the "creator", a term used by some Indigenous peoples. He, as with the other interviewed pre-service teachers, was careful, thoughtful, and humble about his knowledge of Indigenous perspectives. However, he also feared appearing tokenistic.

Learning to Value the Connections Amongst Entities in the Garden. All three participants learned to value the connections amongst entities in the garden. Participant 1 was eloquent about the value of connections, commenting on their importance to Indigenous peoples by saying:

> Well a lot of the Indigenous features are focused around things on the earth, so animals and the sun, basically Mother Nature, and from what I know now of Indigenous perspectives, I still don't know a lot, but I know a little bit more than when I first started is that they have a very strong connection with the natural world, like everything … I mean they lived on and off the earth, right, and just like I guess we all do, but in a very different way, in a spiritual way, and a lot of the ceremonies or everything about their ways of knowing and ways of life have been from the earth.

She went on to elaborate about the buffalo bean, a plant that flowers just before the bison arrive (in the past). Indigenous people paid attention to these details, so they knew what to expect. She, herself, had a strong connection to plants, and through that initial plant connection, came to develop greater appreciation and understanding of Indigenous relationality to all entities.

Participant 2 recognized that connections were inherent to her way of being in the world, noting: "*everything is connected. And I think even just treating one thing poorly is going to have a huge effect on everything that happens to everyone in the future*". Participant 3 said:

> Connections between the garden and history, etc.: the Earth turtle and the creation story. The cooperation and interconnectedness between all the species, different animals working together to contribute to the well-being and survival of the falling woman [Sky Woman] and I believe this is analogous to our relationship to other beings on the earth. A mutual and cooperative and respectful relationship. They help us a lot and we need to help them. Reciprocation in that regard too, would be another key word.

These two summer students were applying relational concepts to their material experiences of the garden. Their focus, in each case, was on the relationships—that the falling woman from the creation story of Turtle Island[3] was supported by the other animals, and that everything is connected.

The material reality of the garden was integral to the development of understanding of the value of interconnections and relationships. Interestingly, all three of these pre-service teachers spoke tentatively about Indigenous perspectives, despite having read much material by Indigenous peoples from the great plains of North America. They recognized that they are of settler ancestry and must not claim ownership of this knowledge. Also, they recognized that their knowledge was incomplete.

Participant 1 commented on the change for her, from gardening with intention to produce efficiently what she wanted, to recognizing the value of all entities as contributors to the garden:

> I remember the first time I went in there and I was going to get rid of all the leaves and clean up the yard like you would at home. But then I realised but wait, we need those leaves because they are a natural mulch, like I didn't know that, because I am used to seeing the soil and spraying herbicides to get rid of the weeds. These are the things that I knew about gardening but that is not the only way to do it, that there are much more sustainable ways to do it, like where you are re-using materials that are already there for the same thing, without affecting. And growing appreciation for those weeds. So I also learned those things that I would like to use in my own garden eventually.

Indeed, the leaf litter creates homes for fungi and bacteria that add to the health of the soil. Fungi communicate messages from the roots of one tree to the roots of another (Wohlleben, 2016), all supporting the relationality within the garden: the biotic, the soil, air, water, sun.

Participant 2 noted how all things affect all other things. She had been aware of this as a child, but the material realities of having to earn a living to pay her way through her university education separated her from this thinking. Being in the garden returned her to understanding the value and care she had for all other beings. She noted learning about the connections that plants make with one another, and then went on to the connections she made to one of the First Nations Elders we worked with:

> I guess it's just that everything in the garden was together, and different plants were helping with other plants, like certain chemicals that they release into the ground and something that you told me about, that help or hinder another plant, so it's just interesting – going to use the same word again – but how interconnected everything is. And just how all of the plants and the structures in the garden are together, and I guess just all, like some of the things that [our Elder] made in the garden are hard to notice. It's not taking attention away from any other part of the garden.

Everything is connected, everything is valued and valuable.

Participant 3 started his interview by noting the valuing of connections to Other, and that this was the first experience he had in the garden, during an undergraduate course: "*I remember feeding the birds, and that set the tone for interaction and interconnectedness with other species and the environment, and interconnectedness between living and non-living parts of the environment as well*". The material reality of the birds, of that connection between the bird and himself, led him to value that bird, but also to value the ecosystem.

Participant 1 stated:

> And a lot of it unfortunately was selfish, it was for me. I mean I did help the garden, but the garden helped me too. It helped me get to a point where I can use a lot of the information that I did gather from there to kind of like grow my passion and hopefully to promote the importance of the native plants in Saskatchewan, the prairies and also our relationships with Indigenous people. So those are two things that I didn't realize were connected to my own passions of being outside that I realized "oh my goodness, this is all one, together" if that makes sense. It was almost therapeutic and transformational. It was just a wonderful time. I don't know how else to describe it.

Relationality is the main concept worked and discussed, according to them. They revealed the concept in a set of analogies/metaphors to demonstrate the valuing of *self and identity in the context of community*; as a way to value *cooperation, interconnectedness* with Others (as person, entities, or cultures), and a *set of relationships that make ecosystems*, supporting a biodiverse and healthy planet.

But how do teachers and pre-service teachers of settler ancestry plan to take Indigenous perspectives/relational ontologies into classrooms?

Learning to Value and Teach Indigenous Perspectives. It is integral to note again that there is not just one Indigenous perspective; there are multiple. Indeed, if we get into details, every person has a unique perspective. However, for the purposes of this chapter, and for the participants' initial teaching experiences, they had learned about the main aspects of Indigenous ontology from multiple local Indigenous people's perspectives. The academics they read, the academics they talked to, the Elders who supported their work in the garden, presented a relational ontology, focusing on relationships amongst entities, as well as sovereignty issues—who owns the land, and who owns the knowing of the land.

The three participants who had worked in the garden for the summer, and who were interviewed, all expressed humility regarding Indigenous knowing and perspectives. All three pre-service teachers identified as white; two were from backgrounds that would be considered "white settler". Thus, they believed they did not know, and certainly could not claim as their own, Indigenous perspectives. Participant 1 noted "*I still don't know a lot, but I know a little bit more than when I first started*". Similarly, Participant 2 noted: "*I am not an expert or anything.*" Participant 3's humility emerged in terms of his concern regarding tokenism. "*Also very spiritual, not an idea of God so much, but I developed a greater connection to – not the creator really, but to creation. In relation to my settler identity. Is this tokenism? A strong sense of connection to creation.*" He also addressed the concept of sovereignty, in stating that he no longer considered that he could own land, but that rather, we belonged to the land, and that we had a duty to care for it, and to learn from it. Reforming the DWW concept of ownership and entitlement is integral toward reconciliation.

It is important that settlers retain uncertainty and humility when addressing Indigenous perspectives. Remaining humble about what one knows can make one more open to learning. Coming to understand one's own, let alone another culture's, perspective is a life-long process.

All three participants described how working in the garden supported them developing more understanding of Indigenous history, culture, and relational ontology. A starting point is necessary, and these pre-service teachers described how they took this information into their teaching practice. These teachers, in their practice, knew enough to steer their students in the right direction for their learning, but also the teachers knew they could contact Indigenous Elders, or knowledge keepers.

For example, they addressed the knowledge that had been developed, such as that represented in the Celestial Circle, but also the attitudes towards other species and relationships. They also noted the role of ceremony. They had all worked with at least one, sometimes two Indigenous Elders, so knew that there were resources in the community to draw on. They had learned that there are protocols for working with Elders, and that there are some protocols that can suffice but that each Elder will honor specific details.

Participant 1 stated:

> That student [graduate student, who installed a replica] was doing a project on the Celestial Circles, I think, or something like that, and that was really interesting actually. I enjoyed how she had done a whole ceremony around it, but yeah it was really neat to be part of that process of creating it with the community with everyone else there. There was also the sun effigy I think that we did with JN [Elder].

Participant 2 was able to introduce her students to the garden; she wanted them to use the garden as inspiration for studying Indigenous art and symbols. She knew enough that she could suggest what might be included in student art-work, but also, she could see when they needed to do more research:

> They even went into depth like for the symbolism for certain things for Plains Cree groups because that would have been the group that would have been here how many years ago as well. And some of them were starting off with symbols from other groups, and I was able to kind of steer them in the right direction, and even just the colours around the Medicine Wheel that they were drawing, I was able to get them to look up and make sure that what they were drawing actually represented the people that lived here and was respectful of that.

Participant 2 tried to get an Indigenous knowledge keeper to come to her classroom, but provincial cutbacks to education meant that there were insufficient school division resources to bring guests to her classroom.

Interestingly, although Participant 2 noted the Celestial Circle, and the role it played for Indigenous peoples, when her high school students insisted on interpreting it as a Medicine Wheel, she did not stop them. She attempted to guide them, but they were resistant—they were perhaps too confident in their beliefs about Medicine Wheels to be able to imagine other kinds of circles on the plains. Somehow, teachers will have to ensure that their students of settler ancestry do not, in their learning about Indigenous history and perspectives, become so "knowledgeable" that they appropriate this knowledge, and speak on behalf of Indigenous peoples.

Participant 3 noted about his learning:

> consideration of our Indigenous history, demographics, traditional worldviews, and knowledge. Connections between the garden and history, etc.: the Earth turtle and the creation story. The cooperation and interconnectedness between all the species, different animals working together to contribute to the wellbeing and survival of the falling woman and I believe this is analogous to our relationship to other beings on the earth. A mutual and cooperative and respectful relationship.

He itemized all the different Indigenous features in the garden and noted how each represented an aspect of Indigenous perspective. About the bison, he said:

> And then, as well, it evokes the notion of the settlers' extinction of the buffalo which reflects their world view and representative of that concept of sustainability, and it is a tangible representation of that lesson so relevant in today's age.

Having some understanding of Indigenous history and perspectives, having worked with Elders, and having participated in ceremonies (including appropriate protocol for working with Elders, smudging, and building replicas of

Indigenous artefacts), supported these pre-service teachers as starting points in their teaching. After one (Participant 3), two (Participant 2), and three (Participant 1) years of teaching, they retained their humility, staying open to learning more, continuing to read works written by Indigenous peoples, and through working with Indigenous Elders, knowledge keepers, and community members.

All three pre-service teachers noted how they went about integrating Indigenous history and perspectives into their teaching. For all of them, they approached new learning through the subject areas they were comfortable with. Participants 1 and 3 taught elementary school, and so were expected to teach all subject areas. Both had backgrounds in science, so they began there, and found easy connections for including social studies, history, and Indigenous perspectives. Participant 1 gave an example from her teaching, beginning with science inquiry exploring native prairie, and drawing in the social studies action project (asked for in the curriculum), to include Indigenous history, and relational ways of being, with the relationality emerging through action with the prairies. She said: "*These things were already known, and they were already here. We just need to learn about them. And so that is kind of how I go about teaching those concepts.*"

Participant 2 was a high school art teacher. She used art to connect to relationships with other species. She had her students research the topics connected to the garden, and most of them chose to represent Indigenous perspectives. When students seemed to be taking up a naïve approach, she had sufficient knowledge, for the most part, to encourage them to do more research. There were still examples where her high school students had been unable to move to the depth of knowledge that she had, but since she remained humble about her understanding, she was not harsh with them. She encouraged them to remain open to learning more. The question for her emerged about her students appropriating Indigenous symbols in their art. However, since this was a learning project, she was comfortable. The students were not creating work to sell but were representing their learning of art and of Indigenous history and perspectives.

Another strategy that one participant picked up on was to use some of the programs available through his school division, such as the grandfather teachings and the seven generations concept. He hoped he was modeling humility, for example, and courage—two of the grandfather teachings.

Each of the participants started in a place of comfort, where they felt they knew the topic (for two, this was science, for one was art, and participant one also drew on early learning inquiry/play/story-telling) and used this as their way to flow into Indigenous history and perspectives, about which they remained humble. They found that connecting Indigenous history and perspectives to their areas, although still rife with tensions, was easier than stepping into unknown territory. One of the participants had her students carry out their own research and used her understandings to encourage them to go deeper to get greater accuracy. Another one drew on school programs, such as the grandfather teachings and the seventh generation concept, to support his teaching. Nonetheless, the tensions must be embraced, as we learn more about what to do, and when.

Discussion

In the generation of theory, we were predisposed to look for whether the pre-service teachers were developing ideas of relationality. In this search, we learned a number of aspects that we believe are important and worthy of more thorough investigation. One is that sustained involvement in a naturalized place makes a difference. The data also reveal the importance to spend time alone as a strategy to create personal engagement and acknowledgment of one's own body and self in the environment.

The summer-student pre-service teachers found that they learned to value themselves, through connecting to the other aspects of the garden. Thus, the kind of mindfulness where one becomes aware of self as part of a system is something that should be explored in more detail.

All three of the pre-service students who were interviewed noted the physical materiality of the garden. It was not just about the mind, or even only about the emotions, but rather a total embodied connectedness to the land. Having their hands in the earth, listening to the birds and wind, seeing the flowers and the species that interacted with them, all this affected the summer students in learning to value themselves, connections, and the ecosystem.

One valuable lesson emerged: the importance of place. Sustained involvement in this naturalized place, connection to Indigenous Elders and readings, supported the teacher candidates in focusing on teaching as caring, values, and relationships. As well, the pre-service teachers stated that they themselves are valuable in the relationships they formed with all other entities—including winter weather, rocks, and those annoying little ticks.

Noun-based languages, in this case English, is dominant in this place, and thus places greater value on entities. Entities do exist; they are valuable. This valuing of entities derives from or supports DWW ontologies and epistemologies, as pointed out by Seawright (2014). However, we can escape the trap of our language through using nouns that address relationships; what remains is to overcome our belief that only individuals, and those specific individuals are all human (and mostly white DWW male humans), are the entities that matter. The Indigenous peoples of the great plains of North America, according to their scholars (for example, Cajete, 2005; Deloria, 2001; McGregor, 2005; Simpson, 2011, 2014), valued relationships. It is not that they did not also see and value entities, but that far greater value was placed on the relationships. No entity was capable of existing, except in relationship with other individuals of the same species, other species, and with the land, air, water, and so on.

It would seem that children are capable of focusing back and forth, between entities and relationships (McVittie, Datta, Kayira, & Anderson, 2019). It is through interactions with adults that they come to see the world as a set of individual entities, through survival demands in this individualized world, and then go on to consume material goods as if these will satiate them. And yet, pre-service teachers, having grown up in a society that values individuals, and individual entities (generally), are still capable of changing their ontologies. This was seen when they had sustained experiences with the PHG, where they spent time with their hands in the soil, nurturing plants, noticing invertebrates, and other species, and thinking about Indigenous features. They shifted their perspectives in ways that one participant called "transformational". They no longer looked just at a yellow spider but saw it in relationship to the yellow cone-flower and to the unsuspecting bee coming to pollinate the flower. Similarly, some baby warblers grow up to be adult warblers, and some grow up to be crow food, and our interactions through removing wild roses put the warbler babies at risk but offered easier food for the baby crows. Thus, it is by re-educating our attention (Ingold, 2011) that we become actively aware of our involvement in the web of relationships.

Language matters in epistemologies, affecting what is believed to matter, to count. From this research, it would seem that material also matters; the material experience of putting their hands in the ground made a difference to the pre-service teachers. Over the sixteen weeks of their summer roles, the pre-service teachers embraced the time to be with the material, physical world, a time for them to be calm and to think. Time alone, with the physical material world is required to develop relationships with entities in places, and with the places themselves.

Interestingly, because the values that were promoted in the PHG derived from work with Indigenous elders, and from reading Indigenous scholars from the North American great plains, the pre-service teachers were reluctant to claim these perspectives. However, they lived these perspectives, talked and taught them. Teachers of settler ancestry should worry about appropriating

Indigenous perspectives and ways of knowing, but if we are to change our teaching from a DWW ontology that foregrounds entities to an Indigenous ontology that foregrounds relationships, we must acknowledge Indigenous peoples; they must be honored for their contribution to a larger philosophy. The humility with which the summer students spoke about Indigenous perspectives (indeed, about their own perspectives) was important, and they had various strategies for not taking on the expert role. They knew enough to feel comfortable inviting Elders and knowledge keepers to their classrooms (although budget cutbacks meant they had insufficient access to Elders and knowledge keepers), could connect their subject areas of interest seamlessly with Indigenous history and perspectives, and could support their students in carrying out their own research.

These pre-service teachers honored the people who have lived a relational ontology; as well, the pre-service teachers have taken it up in their own life philosophies. In a variety of ways, they said *"I learned this, from listening to Elders, in reading the works of Indigenous scholars, in paying attention to the land. This is now how I live"*.

The theory developed is thus: to learn a relational ontology, there were at least three aspects that we noticed: (1) the pre-service teachers were involved for a sustained period of time in a material place, with a physical involvement in that place; (2) the pre-service teachers had much solo and meditative time to reflect on the place, the land, the relationships; and (3) the pre-service teachers had read and discussed ideas with Indigenous scholars and Elders, and thus were primed to examine the land in relational ways. These led them to value themselves as they come to understand themselves in relation to their history and the land; to value the connections amongst all the entities in the garden; and to value and to teach Indigenous perspectives of relationality. If these values remain, they will live in respectful ways with all entities, not prioritizing their own wants above those of the earth.

We wonder about the lasting effects of the PHG on these pre-service teachers as they go through their teaching careers. Will the children who learn with these teachers grow up in respectful, caring, relational ways; will they request to hear a variety of stories rather than single stories; will they refuse to hear stories which demand they selfishly seek more "stuff" for themselves; will they consider how their "owning" of property is immoral, unrealistic, and that consumption will, immediately and in the long term, damage Others? Through dwelling in their cityscapes, will they come to understand urban areas as relational; will they notice and value the relationships amongst entities in that place? Indigenous peoples lived on and with the land, and this was sustained—not just for six hours a day, for sixteen weeks, but rather over their lives, and their families lived there over millennia. Indigenous peoples learned their roles, and the roles of other entities. Our roles as humans, as they noted, are to come to understand Other, to recognize and value the interconnections, to know the fabric of life that maintains us all. It is never about one truth; it is always about the ability to pay attention to Other, so as to learn to value those relationships that allow us to exist at all.

We also remain in tension regarding the choices that Indigenous peoples will make regarding the land. Settlers have done great damage to the land, to ecosystems, and have created a mass extinction crisis. At the same time, settlers have marginalized the original peoples who lived here sustainably for milennia, with settlers attempting genocide, and leaving survivors in poverty. It is time to ensure that Indigenous peoples manage the land. If the leaders of the Indigenous peoples choose to extract resources in damaging ways, settlers are in no position to criticize. However, we are hopeful because of the research that so far demonstrates that greater biodiversity exists in Indigenous managed lands.

Notes

1. "Other" is capitalized when used as a noun.
2. Deloria here refers to the peoples now named 'American Natives'. In Canada, these people have chosen to name themselves 'First Nations or Métis or Inuit' peoples; in showing relationship to First Peoples globally, the term 'Indigenous' is often used.
3. Turtle Island derives from the story of Sky Woman falling from above, landing in the water, and getting on to a turtle's back. The story is known by almost all North American Indigenous peoples, regardless of whether the story is part of their ancestral stories.

References

Alexis, T., & Poscente, P. (2019). Opinion piece, Globe and Mail Newspaper, <https://www.theglobeandmail.com/opinion/article-the-trans-mountain-pipeline-and-indigenous-ownership-is-too/>.

Basso, K. H. (1995). *Wisdom sits in places: Landscape and language among the Western Apache*. Albuquerque, NM: University of New Mexico Press.

Bowers, C. A. (2001). Addressing the double binds in educating for an ecologically sustainable future. *International Journal of Leadership in Education, 4*(1), 87–96.

Bowers, C. A. (2010). Educational reforms that foster ecological intelligence. *Teacher Education Quarterly, 37*(4), 9–31.

Bryant, A., & Charmaz, K. (Eds.). (2010). *The Sage handbook of grounded theory*. Newbury Park, CA: Sage Publishing. <http://dx.doi.org/10.4135/9781848607941>.

Cajete, G. (2005). American Indian epistemologies. *New Directions for Student Services, 109,* 69–78.

Calderon, D. (2014). Speaking back to manifest destinies: A land education-based approach to critical curriculum inquiry. *Environmental Education Research, 20*(1), 24–36.

Clarkson, L., Morrrissette, V. & Regallet. G. (1992). *Our responsibility to the Seventh Generation: Indigenous peoples and sustainable development* (p. 88). Winnipeg, MB: International Institute for Sustainable Development.

Coulthard, G. (2010). Place against empire: Understanding Indigenous anti-colonialism. *Affinities: A Journal of Radical Theory, Culture, and Action, 4*(2), 79–83.

Deloria, V. Jr. (2001). American Indian metaphysics. In V. Deloria, Jr., & D. Wildcat (Eds.), *Power and place: Indian education in America* (pp. 2–8). Boulder, CO: Golden Press.

Doherty, B. (2019). Queensland extinguishes native title over Indigenous land to make way for Adani coalmine. *The Guardian.* <https://www.theguardian.com/business/2019/aug/31/queensland-extinguishes-native-title-over-indigenous-land-to-make-way-for-adani-coalmine>.

Ingold, T. (2011). Against space: Place, movement, knowledge. Chapter 2 in P. Kirby (Ed.), *Boundless worlds: An anthropological approach to movement* (pp. 29–43). Oxford, UK: Berghahn Books.

Intergovernmental Panel on Climate Change. (2018). Global warming of 1.5°C. An IPCC special report. <https://www.ipcc.ch/sr15/>.

Kenny, M., & Fourie, R. (2014). Tracing the history of grounded theory methodology: From formation to fragmentation. *The Qualitative Report, 19*(Article 103), 1–9.

Kenny, M., & Fourie, R. (2015). Contrasting classic, Straussian, and constructivist grounded theory: Methodological and philosophical conflicts. *The Qualitative Report, 20*(8), 1270–1289.

Kimmerer, R. W. (2013). *Braiding sweetgrass: Indigenous wisdom, scientific knowledge and the teachings of plants.* Minneapolis, MN: Milkweed.

King, N. (2017). *Medicine wheels and celestial circles: History, symbolism and teachings.* Unpublished Masters project, School of Environment and Sustainability, University of Saskatchewan.

McGregor, C. (2005). Traditional ecological knowledge: An Anishnabe woman's perspective. *Atlantis, 29*(2), 103–109.

McVittie, J., Datta, R., Kayira, J., & Anderson, V. (2019). Relationality and decolonization in children and youth garden spaces. *Australian Journal of Environmental Education, 35*(2), 93–109.

Patel, R. (2009). *The value of nothing: Why everything costs so much more than we think.* Toronto: HarperCollins Publishers Inc.

Renwick, A. R., Robinson, C. J., Garnett, S. T., Leiper, I., Possingham, H. P., & Carwardine, J. (2017). Mapping Indigenous land management for threatened species conservation: An Australian case-study. *PLoS ONE, 12*(3), 1–11. e0173876. <https://doi.org/10.1371/journal.pone.0173876>.

Ritchie, J. (2015). Food reciprocity and sustainability in early childhood care and education in Aotearoa New Zealand. *Australian Journal of Environmental Education, 31*(1), 74–85.

Schultz, R., Abbott, T., Yamaguchi, J., & Cairney, S. (2019). Australian Indigenous land management, ecological knowledge and languages for conservation. *EcoHealth, 16,* 171–176. <https://doi.org/10.1007/s10393-018-1380-z>.

Schuster, R., Germain, R. R., Bennett, J. R., Reo, N. J., & Arcese, P. (2019). Vertebrate biodiversity on indigenous-managed lands in Australia, Brazil, and Canada equals that in protected areas. *Environmental Science & Policy, 10*(1), 1–6.

Seawright, G. (2014). Settler traditions of place: Making explicit the epistemological legacy of white supremacy and settler colonialism for place-based education. *Educational Studies, 50*(6), 554–572. <https://doi.org/10.1080/00131946.2014.965938>.

Seldén, L. (2005). On grounded theory – With some malice. *Journal of Documentation, 61*(1), 114–129.

Simpson, L. B. (2011). *Dancing on our turtle's back: Stories of Nishnaabeg re-creation, resurgence, and a new emergence.* Winnipeg, MB: ARP Books.

Simpson, L. B. (2014). Land as pedagogy: Nishnaabeg intelligence and rebellious transformation. *Decolonization: Indigeneity, Education & Society, 3*(3), 1–25.

Truth and Reconciliation Commission of Canada. (2015). *Truth and Reconciliation Commission Calls to Action.* <http://nctr.ca/assets/reports/Calls_to_Action_English2.pdf>.

Wildcat, D. (2001). Indigenizing education: Playing to our strengths. In V. Deloria & D. Wildcat (Eds.), *Power and place: Indian education in America* (pp. 7–19). Golden, CO: Fulcrum Publishing.

Wildcat, M., McDonald, M., Irlbacher-Fox, S., & Coulthard, G. (2014). Learning from the land: Indigenous land based pedagogy and decolonization. *Decolonization: Indigeneity, Education & Society, 3*(3), I–XV.

Wohlleben, P. (2016). *The hidden life of trees: What they feel, how they communicate; Discoveries from a secret world.* Vancouver, BC: Greystone Books.

World Wildlife Federation. (2018). *How many species are we losing?* <http://wwf.panda.org/our_work/biodiversity/biodiversity/>.

8. Environmental Education through Indigenous Land-Based Learning

Ranjan Datta

from the Laitu Khyeng Indigenous at the Chittagaong Hill Tracts (CHT), Bangladesh.

Introduction

In this chapter, I share my learning reflections on the meanings of environmental education (EE) from the Indigenous land-based knowledge of Laitu Khyeng Indigenous Elders and Knowledge-keepers at the Chittagong Hill Tracts (CHT), Bangladesh. Following an Indigenous relational theoretical framework, I share my learning reflections about how members of the community understand the meanings of EE from their everyday Indigenous land-based knowledge and practice, including what and why Indigenous land-based knowledge is important for understanding and practicing decolonization in everyday life? How does taking an Indigenous land-based approach support advocacy for Indigenous rights? How do we know it is working?

A relational theoretical approach in EE is deeply connected with Indigenous land-based learning (Bang et al., 2014; Cajete, 1994, 2016; Calderon, 2014; Datta, 2018; Goeman, 2008; Grande, 2014; Simpson 2014, 2017; Styres, 2019) but cannot adequately be covered in this chapter due to the limitations of space. I highlight my specific learnings from the Laitu Khyeng Indigenous community to demonstrate their Indigenous relational theoretical perspective. I worked and learned with the Laitu Khyeng Indigenous community, and in this chapter, I discuss my lessons learned. It is also important to note that Indigenous identity and traditional knowledge as considered in CHT are neither static nor linear, but rather hybrid, fluid, and varying from community to community, land to land, and generation to generation (Datta, 2018). Indigenous identity and traditional knowledge are sacred and living ways of thought and action rooted in the lands where Indigenous people live. In my decolonial learning journey, I learned that the Laitu Khyeng define and identify traditional knowledge using diverse relational ontologies, epistemologies, methodologies, and also have diverse pedagogical ways of passing it on to their younger generation.

For achieving my goal of understanding how an Indigenous land-based knowledge can enrich EE within and from our everyday practice, I have divided this chapter into five sections: first, I situate my research framework, methodology, and methods; second, I share my learning experience about the meanings of land-based learning from the community Elders and Knowledge-keepers' stories; third, I discuss how land-based learning helped me to understand the colonial process; fourth, I reflect on how land-based learning can promote Indigenous rights; finally, I evaluate how we can implement Indigenous land-based learning in our EE, through rethinking EE from land-based practices, developing our critical lens, and advocating for Indigenous rights.

Research Framework, Methodology, and Methods

In exploring an Indigenous land-based approach from an Indigenous community perspective, I use a relational theoretical framework, as it "not only challenges Western fixed meanings of actors but also makes actors responsible for their actions" (Datta, 2015, p. 102). An Indigenous land-based framework considers "multiple realities, relationships, and interactions based on our traditional knowledge" (Datta, 2015, p. 107; see also Bang & Marin, 2015; Cajete, 1994, 2016; Simpson, 2014, 2017; Wilson, 2008). In a relational framework we, as human and non-human, living and non-living, become responsible to each other for our understandings and actions. Moreover, an Indigenous land-based framework demonstrates the benefits of using Indigenous ways of knowing and doing. My learning reflection is also part of Participatory Action Research (PAR) as it transformed me into who I am as a researcher and who I should be as a researcher in the community (Datta, 2018). According to the Elders and Knowledge Keepers in this study, sharing my learning is also part of my relational responsibility, as my learning helps to reshape who I am as a researcher (Datta et al., 2015; Wilson, 2008).

There are many themes discussed in this research and one of the themes is land-based education. In this chapter, I endeavor to explore my learning experience with the Laitu Khyeng Indigenous Elders and Knowledge-keepers regarding how the community envisions Indigenous land-based learning. Elders are recognized as respected knowledgeable elderly people in the community who can guide the community. A Knowledge-keeper holds significant traditional knowledge for the community. In the community, a Knowledge-keeper (as identified by the community) could be both Knowledge-keeper and Elder; on the other hand, an Elder may not be identified by the community as a Knowledge-keeper. Both Knowledge-keeper and Elder are respectful to the community. Elders provide their guidance through their knowledge and wisdom and Knowledge-keepers hold important knowledge on culture, tradition, and well-being. I developed my learning reflections from three research methods: (1) Elders and Knowledge-keepers' collective traditional story sharing (we had a total of five traditional story sharing events during my six months of field research, including beginning of field research for exploring research objects, during field research, after field research data collection, after collective data analysis, and during research sharing); (2) individual story sharing with the Elders and Knowledge-keepers (we had total fifteen story sharing events from fifteen Elders and Knowledge-keepers); and (3) my personal learning reflections derived from my notes. Most of the quotations in this chapter were transcribed and translated in a collaborative process with four Khyeng Indigenous co-researchers from the community and myself as an academic researcher. Elders and Knowledge-keepers reviewed transcriptions; however, once I had done the second step in translating to English, the Elders and Knowledge-keepers, who do not speak English, could not review them. For protecting Indigenous Elders' and Knowledge-keepers' identities, as was their choice, I did not use names or pseudonyms, as they believed it would be risky to disclose who they were. Since Khyeng Indigenous people's names are culturally connected and since it is a small community, it would have been easy to identity them, even if I had used pseudonyms. Therefore, to protect Elders' and Knowledge-keepers' identities, I did not use pseudonyms for their quotations, but rather amalgamated the data.

Following the Indigenous community's protocols and ceremonies, I explore my learning experience for two main goals: (1) sharing my learning experience from the Laitu Khyeng Indigenous community Elders and Knowledge-keepers regarding how they suggest the meanings of their Indigenous land-based approach from their perspectives; and (2) explaining why and how the Laitu Khyeng Indigenous community Elders and Knowledge-keepers think the Indigenous land-based approach is not only useful for learning how to decolonize, but also for reclaiming Indigenous research and ways of knowing and for advocating Indigenous rights.

Land-Based Learning and Environmental Education

Through coming to understand their land, their traditional knowledge, and their rights, they worked towards decolonization. In other words, the Elders and Knowledge-keepers viewed EE as a process that is lived on the land, and that it happens informally, in the community, rather than in the school. I came to understand how these diverse meanings and practices are connected with their Indigenous identity and with everyday practices.

Using a traditional story sharing conversation between myself and a group of Elders and Knowledge-keepers, a Knowledge-keeper explained "The meaning of the land-based learning for our community is everything, including our land, water, jungle rights, our traditional knowledge, our Elders and Knowledge-keepers, and our foods." This Elder also said, "Our meanings of land significantly differ from mainstream non-Indigenous community (i.e., mostly for Muslim people). In the mainstream, people call land an asset—such as my wealth, my home, my land, my cattle. However, Indigenous people understand wealth as relationships, a collective responsibility". Another Elder also explained land-based learning in a similar way: "We believe that we belong to nature, not that nature belongs to us. The sun is for everyone and the wind is for everyone, as the water of the river is for everyone, so is the land".

Thus, I came to understand that ownership was relational for the Indigenous peoples. There was no private ownership, nor public, nor was there a cooperative organization. Rather, they collectively lived in relationship with the land, and with all entities on the land.

Another Elder explained that the philosophy of the Indigenous land-based learning is their everyday practice. He said:

> The philosophy of life of the Indigenous people is how I can leave what I have found, what my nature has given me or us, to my next generation - not just my children, my grandchildren, and the next generation. Moreover, it is not just that we talk about our people; we think about collective ways of living. We do not refer to the collective as only for humans. Our ways of doing things connect everything, including humans, animals, plants, birds, water. We believe the tree has a life, has the power to provide us with food, and protect us; we as humans have the responsibility to protect them. Therefore, we want to keep everything for everyone in the larger world, the human beings, the plants, the animals together.

From this quote I learned, the community sees the importance of both learning human and nonhuman in Indigenous land-based learning.

Knowing Indigenous identity from an Indigenous land-based approach is not only helpful in redefining the meanings of land, but also it helps to reclaim meanings of Indigenous identity within and from their land-based practice. For instance, one of the Elders said "We need to know the identity that we can belong to. How did we grow up and connect in our land? Our language, culture, songs, stories are still alive. We need to connect with land, ceremonies, spiritualities, animals, plants". The Elder, in this way, demonstrated that an Indigenous land-based learning is their philosophy.

Another Knowledge-keeper suggested that an Indigenous land-based approach involves ways of knowing and doing:

> Indigenous land-based approach is the way I have found my life and the world so that I can leave it for my next generation. That is Sustainability. I can live on my own, and I will not eat everything. I will use it in my life's needs but will develop and do it again. So that my next generation can adopt it and further enhance it. That is the Indigenous land-based approach.

An Indigenous land-based approach as traditional knowledge can create challenges to colonial approaches. For instance, during an individual story sharing, a Knowledge-keeper stated

> Our traditional knowledge is a challenge to the government's colonial system. The government system stands on the personal profit system. Their [Bangladesh government] system is essentially anti-Indigenous. Therefore, their system is to destroy Indigenous collective and holistic ways of understanding and practicing. In the government system, everything counts as money, everything is considered as profit. Therefore, our people and our traditional knowledge became the target of the Bangladesh government colonial system. Therefore, outsiders [mainstream settlers] came with logging; land ownership, colonial religion [Islam] to grab our land. They [mainstream settlers] came with only ownership of the land and the possession. Their colonial understanding and practice became oppression to our women and the environment. Their [mainstream settlers] come with oppression, arms trade, killing, raping Indigenous women, and profit maximization. Reclaiming our traditional knowledge is a challenge for the mainstream colonial system.

Indigenous land-based learning refers to relational responsibility. For instance, Elders and Knowledge-keepers during our conversation suggested that in a land-based approach it is everyone's responsibility to protect our land, water, forest, and animals. Responsibility is defined as protecting land, water, Indigenous identity, traditional knowledge, and culture.

As I listened to the Elders' and Knowledge-keepers' perspectives on land, I began to recognize how their sharing of their stories with one another added complexity and depth to each story. I realized that Indigenous land-based learning is a lifelong process. Indigenous beliefs about land are connected with their everyday practice, particularly how to be responsible to the land and relationships. Indigenous land-based learnings have always been grounded in the land, mobilized by cultural practices and teachings that shape our understanding of the world and our responsibilities within it. Therefore, the land-based education in Indigenous communities can create many opportunities for both Indigenous and non-Indigenous people to consider linking both educational policies and diverse practices that support Indigenous land-based education with community capacity-building actions.

Coming to Understand the Colonial Process and Decolonization

Indigenous land-based learning plays an important role in decolonization by challenging colonial histories of the land. I learned that the colonial system is the root of most of the challenges to the Indigenous people for engaging in their traditional EE in CHT. For instance, a Knowledge-keeper explained how Indigenous land-based learning can help us to decolonize our ways of knowing and doing. He argued that the colonial system will not be destroyed by the colonizers (i.e., mainstream), since they rely on the colonial system to maintain their power and wealth. Other Knowledge-keepers also provide similar examples that point to the way the Bangladeshi government and non-governmental agencies have created many projects for displacing Indigenous people from their land. For instance, he said

> [There are] 500 military camps in our lands, many lumber plantation projects are cutting our natural forests, there are many national and multinational agencies, tobacco, gas, and oil companies. In 95–99% cases, Indigenous people have not been involved with these exploitative projects. These projects were created by the outsiders, for the outsiders, for grabbing Indigenous land. You can find all this information from many prior research studies. I think, Bangladesh government will not destroy their colonial system as they are benefitting from this system.

Another Indigenous Elder explained that land-based learning helps us to know the colonial histories of the land, such as, why the colonial system was created through Bangladesh's government and non-government colonial settlement projects. Similarly, Indigenous scholars Chakma (2010) and Roy (2000) explained how Bangladesh government settlement policies supported the colonial process in CHT. An important factor discussed by many Elders is the Bangladeshi military camps on Indigenous land. For instance, an Elder explicitly stated

> There is the presence of the Bangladeshi army. The military camps have been seriously disrupting the traditional Indigenous life, ceremonies, customs, and culture for a few decades. At the same time, sexual harassment is happening everywhere. The Army has acquired 75 686 acres of Indigenous land in

Bandarban [name of the Indigenous district area] in the hills. They occupied our land without consultation with the Indigenous leadership and the regional council of the CHT. The Bangladesh state is taking away traditional Indigenous land through its various tactics, military, district administration, rubber cultivation, commercial farming, and social forestry.

Knowing how the colonial system has been challenging for their community is discussed as a significant part of decolonization (Adna, 2004; Chakma, 2010; Roy, 2000). For example, a Knowledge-keeper described that the Bangladeshi colonizers destroyed the traditional system by imposing the Bangladeshi mainstream system over the Indigenous system. He said:

> I have seen that the Bangladeshi administration and the military have violated the traditional land rules. Out of the 23 villages, 18 villages became Bangladeshi mainstream-people dominated. Everyday Bangladeshi mainstream people are pushing in, and the Indigenous people are moving to the borderline with limited land.

To decolonize is to learn how mainstream people (i.e., Muslim people in CHT, those who are illegal settlers) created forced displacement of Indigenous people from their ancestors' land (Adnan, 2004; Chakma, 2010). Community stories explain that displacement through illegal settlement in Indigenous land has become a significant challenge to many Indigenous people. For instance, an Elder claimed, "Bangladeshi illegal immigration increased one day by the naming of many unwanted governmental projects such as the Kaptai Electric Dam [a hydroelectric project for outsiders], which displaced more than 100,000 Indigenous people from their land".

In sum, I learned from Indigenous Elders' and Knowledge-keepers' stories that to begin the process of decolonization is to learn the ongoing colonial history of the land and people. Learning colonial stories helps me to understand my responsibility to the land, people, and traditional culture. This learning process also helps me to challenge the current existing injustice process, and to take political positions for the land, people, and traditional land-based education. Through my learning, as Battiste (2013) has pointed out, I can help as a researcher to reclaim Indigenous land rights. I have seen how Indigenous Elders' and Knowledge-keepers' stories help their future generations to decolonize the condition of their land, people, and culture.

Promoting Indigenous Rights

I learned that Indigenous land-based learning promotes Indigenous rights, including Indigenous identity rights in the constitution, traditional knowledge-based learning rights, traditional customary rights, and land rights. Although there is on-going colonization with Indigenous communities in CHT, I learned from Indigenous Elders and Knowledge-keepers that we need to keep building solidarity with each other (including other Indigenous people) to protect Indigenous land rights, Indigenous identity rights, and traditional knowledge rights.

Constitutional Land and Identity Rights

The Indigenous land-based knowledge means constitutional land and identity rights for the community. For instance, a Knowledge-keeper says, "as Indigenous we do not have our constitutional recognition in Bangladesh, but it is the first fundamental right that Indigenous people need". Another Elder explains that constitutional rights are a significant part of their advocacy. She also explained what constitutional rights should look like:

> In Bangladesh, people are from many cultures, many Indigenous tribes, and many languages. This diversity needs to be recognized in the constitution. I think that in the preamble of the constitution, the plurality of Bengal, the plurality of Bangladesh, it should come, and here at least one of the schedules should include names of all nations, names of all languages should be mentioned. The second that I think is to ensure the equal rights of all nations to all races, all languages, all religions. In the sense of ownership, I mean that our constitution has three types of ownerships. It is called private ownership, state ownership, and co-operative ownership. However, the method of owning Indigenous land is through collective ownership. There is no recognition of Collective ownership in our constitution.

Another Elder suggested that collective ownership is part of an Indigenous land-based approach and should be enshrined in constitutional rights:

> We need to amend our constitution to bring about collective ownership. We need collective ownership in land, forests, jungle, and water. Traditionally we had collective ownership in our Indigenous communities; we have it written in our Hill Chittagong Regulation Three. However, it does not have constitutional recognition. We need this recognition in our constitution so that these things can be implemented.

There is a significant need to recognize Indigenous peoples' collective rights to lands, territories and resources in the Bangladeshi constitution. Constitutionally Indigenous land-rights recognition not only contributes to their well-being but also to the greater good, by tackling problems such as climate change and the loss of biodiversity.

Traditional Education Rights

Many Elders and Knowledge-keepers suggest recognition and implementation of traditional educational rights as one of the significant parts of Indigenous land-based learning. For instance, one of the Knowledge-keepers suggested

> Primary education should be ensured in the native language of the Indigenous people. Traditional knowledge should be part of our education. Traditional knowledge also needs to consider a significant part of our education so that our future generation can see its significance. By protecting our traditional knowledge, we should have the opportunity to use other knowledge to find solutions in our everyday issues.

Indigenous traditional education rights is an important part of Indigenous self-determination. Indigenous land-based learning focuses on Indigenous traditional knowledge, practice, and culture. I learned that community Elders' and Knowledge-keepers' knowledge were unconnected with their traditional education.

Traditional Customary Rights

Indigenous land-based knowledge refers to Indigenous traditional customary rights to their land. An Elder explained why Indigenous people need customary rights:

> Traditionally our people [the Indigenous people] did not feel that we need to record the customary practice as we practice every day. As mainstream people have been grabbing our land and forcing their culture on us, now we need recognition of our customary practice.

Similarly, another Knowledge-keeper explained why Indigenous communities need customary rights:

> Some of our lands are in individual ownership; some are collectively owned. The Bengalis [Bangladeshi mainstream people] think that there is much land in the hills [Indigenous land], which has to be occupied. Due to this thought, the Bangladesh government sponsored many thousands of settlers to the Indigenous land from 1975-1985. As a result, there is a demographic change in the entire area. For this government-sponsored illegal migration, many Indigenous people (were) displaced, killed, and became labor in their own land. Terrorism, violence on Indigenous women, and landless people significantly increased. Therefore, we need our customary rights to our cultivation so that we can protect our traditions.

Traditional Customary rights and Indigenous peoples education in Bangladesh are unconnected issues to fill a gap in Indigenous collective knowledge on the traditional customary laws of Indigenous peoples. Moreover, Indigenous customary rights can be the most appropriate way to resolve conflicts and retain social cohesion and unity.

Land and Forest Rights

Land and forest rights are significant parts of the Indigenous land-based approach for many Indigenous communities. Similarly, in this research, many Elders and Knowledge-keepers explained that the Indigenous land-based approach refers to Indigenous land and forest rights for their community. For instance, one of the Elders said

> We, as Khyeng Indigenous people, do not have our rights in our forest and mountains. There were different types of trees here before. They cannot be seen anymore. Earlier, different types of animals were seen in this forest. There were various animals, including tigers, pigs. Since outsiders have been clearing our forest for their lumber plantation over our natural forest, there are no big trees, animals. There are only rubber and teak trees.

Another Knowledge-keeper gave a specific example from his village:

> The hill on which I am now standing is the Langu Hills. Earlier, there were different types of animals, such as deer. We have lost our biodiversity. Outsiders cleaned our forests, planted lumber plantations, and destroyed our biodiversity. We used to hunt animals from our forest. Now the rubber and teak gardens have been planted here.

Land, territories and related resource rights are of fundamental importance to CHT Indigenous peoples since they constitute the basis of their economic livelihood and are the sources of their spiritual, cultural and social identity.

Implications in Environmental Education

Through my learning journey, I learned that the Indigenous land-based learning has many potential implications for EE. For example, explaining Indigenous land-based learning, Indigenous scholars Smith (2012) and Styres (2019) suggest that it is a process, not an event. It is a lifelong process of relational ways of understanding and practicing. It is connected with Indigenous meanings of land, water, traditional knowledge, culture, and western scientific knowledge. In the following I explain how my learning helped me to transform my research into action through taking relational responsibility to relearn the meanings of land-based education and the ongoing colonial process, to reclaim traditional land-based knowledge and practice.

Rethinking the Land-Based Meanings

My learning helped me to rethink the meanings of land-based learning. In land-based learning, we cannot separate one from other things such as land vs water, human vs non-human, nature vs culture. I have seen how land-based learning helps the community to take responsibility for understanding, thinking, and acting. Through talking to the Elders and Knowledge-keepers, I have come to understand that the Indigenous land-based approach in EE is grounded in non-linear perspectives, complexity thinking, and creative inquiry; it shares philosophical principles congruent with Indigenous knowledge systems that are propelling decolonization movements in Indigenous research (Battiste, 2013; Smith, 2012) in many different parts of the world. I have also learned how the Elders and Knowledge-keepers recognize their land-based knowledge as interconnected with Indigenous everyday practice (as does the Maori Indigenous scholar, L.T. Smith, 2012, in New Zealand).

Centering Indigenous Voice

Indigenous land-based knowledge centers Indigenous voice. Smith (2012) suggests that Indigenous land-based knowledge refers to a lifelong decolonizing process for both researchers and participants. She refers to decolonization through Indigenous land-based knowledge as many ways of knowing and doing, such as micro, macro, and midcro (i.e., there are many ways of knowing between micro and macro) processes. I also learned that an Indigenous land-based research from Indigenous perspectives promotes community knowledge and practice. It builds trust in community knowledge and practice and reflects social and community interests. Centering Indigenous voices is critical for Indigenous education. Land-based education centers Indigenous voice, Indigenous practices and culture.

Critical Lens

Indigenous land-based knowledge provides a critical lens to decolonize our ways of knowing and doing (Battiste, 2013; Datta, 2018; Tuck & McKenzie, 2015; Smith, 2012; Wildcat et al., 2014). In explaining decolonization from the perspective of Indigenous land-based knowledge and practice, Wildcat et al. (2014) suggest that decolonization must involve forms of education that reconnect Indigenous peoples to land and to the social relations, knowledge, and languages that arise from the land. They explain that decolonization in Indigenous land-based practice is not linear or fixed, but instead differs in different contexts, places, and methods. Making a similar point, Bang et al.

(2014) center Indigenous land-based education to uncover settler colonialism by redefining EE. In their study, they not only re-center Indigenous land-based perspectives as Indigenous cosmologies, but they also show how Indigenous land-based learning can be useful to reshape us as researchers.

Understanding the Colonial Legacy

The Indigenous land-based approach helps us to understand the legacy of the colonial system. For instance, Smith (2012) explains that the colonial system was created by the colonizers for their benefits, and that this system does not benefit Indigenous people. The colonizers' target is to benefit from the colonial system, maximizing profits for the colonizers. Smith argues that the colonizers have been using Indigenous land, water, and other natural resources for colonizing purposes. She further suggests that the colonial system was and is created to destroy traditional sustainable systems, that since the colonizers destroyed traditional sustainable systems, they cannot rebuild them. The colonizer will not destroy their colonial system as they are standing on it. For achieving decolonization goals, she suggests challenging the colonial system. In this study, I also learned that Elders and Knowledge-keepers' stories helped me to reshape as a researcher, to think and practice land-based learning in our everyday practice. For me, Indigenous land-based learning became a continuous learning process defining who I need to be as a researcher.

Advocates for Indigenous Rights

Indigenous land-based learning promotes Indigenous rights, including constitutional rights of Indigenous identity, land, water, forest, language, and customary practices (Mashford-Pringle & Stewart, 2019; Mowatt et al., 2020). According to Elders and Knowledge-keepers, Indigenous knowledge promotes Indigenous rights, including Indigenous land rights, identity rights, and traditional education rights. They also suggest that knowing Indigenous rights can be helpful in figuring out how to move forward toward environmental sustainability within and with Indigenous communities.

Transforming Both Research and Researcher

My learning opportunity helped me to take responsibility to transform me as a researcher and my research. For instance, in addition to my field research, I had many opportunities to participate as part of solidarity (i.e., Indigenous land-water struggle is my struggle, our collective struggle) in a number of the community's land-right movements (e.g., land-water rights, stopping tobacco growing, stopping brick-field in community); I disseminated research results to local audiences (e.g.: governmental forest, land, and CHT ministries, NGOs, Indigenous research organizations, university professionals, and environmental policy-makers), multinational agencies (UNDP, UNESCO, and Caritas), international seminar presentations (in Canada, Japan, New Zealand, Norway, USA); and published in local and international journals and books. I have come to realize that undertaking this study is a political activity, dedicated to the reclamation of Indigenous and my (as Indigenous and minority) rights (Becker, 1967). Thus, I see that our research is not neutral, rather it is grounded in both an academic and political responsibility to protect and reclaim our rights (which include environmental resource management, sustainability, and identity).

Therefore, my learning reflection from Indigenous land-based knowledge asks how the process of decolonization can be helpful for a researcher (Datta, 2018; Smith, 2012). Battiste (2013) recommends an Indigenous land-based knowledge to analyze education policy, curricula, and pedagogy. The concept of land-based knowledge in EE refers to a lifelong unlearning and relearning process. Unlearning can be taken up by asking critical questions about how western research is not supportive of nor directed through the participants' community; and relearning can be taken up by finding ways to engage the community in the research, how to lead research from within the community, and how to learn from the community. My learning reflections with Indigenous land-based EE point to the way analysis involves the braiding of Indigenous land-based education as diverse knowledge systems to self-determination that is socially just, accountable, and tenable toward a forward vision of generating the most significant potential for others.

References

Adnan, S. (2004). *Migration land alienation and ethnic conflict: Causes of poverty in the Chittagong Hill Tracts of Bangladesh*. Dhaka, Bangladesh: Research & Advisory Services.

Bang, M., Lawrence, C., Adam, K., Ananda, M., Ell, S. S., & George, S. (2014). Muskrat theories, tobacco in the streets and living Chicago as Indigenous lands. *Environmental Education. Research, 20*, 37–55.

Bang, M., & Marin, A. (2015). Nature–culture constructs in science learning: Human/ non-human agency and intentionality. *Journal of Research in Science Teaching, 52*, 530–544.

Battiste, M. M. (2013). *Decolonizing education: Nourishing the learning spirit*. Saskatoon, SK: Purich Publishing Ltd.

Becker, H. (1967). Whose side are we on? *Social Problems, 14*(3), 234–47.

Cajete, G. (1994). *Look to the mountain: An ecology of Indigenous education.* Durango: KivakI Press.

Cajete, G. (2016). *Native science: Natural laws of interdependence.* Santa Fe: Clear Light Publishers.

Calderon, D. (2014). Speaking back to manifest destinies: A land education-based approach to critical curriculum Inquiry. *Environmental Education Research, 20,* 24–36.

Chakma, B. (2010). The post-colonial state and minorities: Ethnocide in the Chittagong Hill Tracts, Bangladesh. *Commonwealth & Comparative Politics, 48*(3), 281–300.

Datta, R. (2015). A relational theoretical framework and meanings of land, nature, and sustainability for research with Indigenous communities. *Local Environment: The International Journal of Justice and Sustainability, 20*(1), 102–113. Retrieved from: <http://dx.doi.org/10.1080/13549839.2013.818957>.

Datta, R. (2018). Rethinking environmental science education from Indigenous knowledge perspectives: An experience with a Dene First Nation community. *Environmental Education Research, 24*(1), 50–66. <https://doi.org/10.1080/13504622.2016.1219980>.

Datta, R., Khyang, U. N., Khyang, H. K. P., Kheyang, H. A. P., Khyang, M. C., & Chapola, J. (2015). Participatory action research and researcher's responsibilities: An experience with Indigenous community. *International Journal of Social Research Methodology.* <https://doi.org/10.1080/13645579.2014.927492>.

Goeman, M. (2008). *From place to territories and back again: Centering storied land in the discussion of Indigenous nation-building* (p. 12). Hanover: Dartmouth College.

Grande, S. (2014). See the forest and think like a mountain. In L. Reynolds (Ed.), *Imagine it better: Visions of what school might be* (pp. 14–20). Portsmouth: Heinemann.

Mashford-Pringle, A., & Stewart, S. L. (2019). *Akiikaa* (it is the land): Exploring land-based experiences with university students in Ontario. *Global Health Promotion, 26*(3_suppl), 64–72. <https://doi.org/10.1177/1757975919828722>.

Mowatt, M., de Finney, S., Wright Cardinal, S., Mowatt, G., Tenning, J., Haiyupis, P., Gilpin, E., Harris, D., MacLeod, A., & Claxton, N. X. (2020). ȻENTOL TŦE TEṈEW (together with the land): Part 1: Indigenous land- and water-based pedagogies. *International Journal of Child, Youth and Family Studies, 11*(3), 12–33. <https://doi.org/10.18357/ijcyfs113202019696>.

Roy, C. R. (2000). *Land rights of the Indigenous peoples of the Chittagong Hill Tracts, Bangladesh.* Copenhagen, Denmark: IWGIA Document No. 99.

Simpson, L. (2014). Land as pedagogy: Nishnaabeg intelligence and rebellious transformation. *Decolonization: Indigeneity, Education & Society, 3,* 1–25.

Simpson, L. (2017). *As we have always done.* Minneapolis: University of Minnesota Press.

Smith, L. (2012). *Decolonizing methodologies: Research and indigenous peoples* (2nd ed.). London, UK: Zed Books.

Styres, S. (2019). Literacies of land: Decolonizing narratives, storying, and literature. In L. Tuhiwai Smith, E. Tuck, & K. Wayne Yang (Eds.), *Indigenous and decolonizing studies in education: Mapping the long view* (pp. 24–37). New York: Routledge.

Tuck, E., & McKenzie, M. (2015). *Place and research: Theory, methodology, and methods.* New York: Routledge.

Wildcat, M., McDonald, M., Irlbacher-Fox, S., & Coulthard, G. (2014). Learning from the land: Indigenous land-based pedagogy and decolonization. *Decolonization: Indigeneity, Education & Society, 3*(3), I–XV.

Wilson, S. (2008). *Research is ceremony: Indigenous research methods.* Halifax, NS: Fernwood Publishing.

Escolas Marginalizadas e Educação Ambiental, *por Julio Karpen, Educador do Campo*
Marginalized Schools and Environmental Education, *by Julio Karpen, Rural Educator*

Às vezes me pergunto sobre o que está acontecendo
As vezes fico com medo do que ainda pode acontecer
Às vezes só ouço lamentos
E novas atitudes nada de acontecer
Sometimes I wonder what's going on
Sometimes I'm scared of what could still happen
Sometimes I just hear laments
And new attitudes nothing to happen

Não paramos nenhum minuto para reflexão
Tantas tragédias vem acontecendo
E tudo se relaciona com a educação e
A falta de de conhecimento
We don't stop for a minute to reflect.
So many tragedies have been happening
And everything is related to education and
lack of knowledge

Deixamos de lado toda preocupação
que envolve o meio ambiente
talvez devido a falta de educação
que o mundo todo está doente
We put aside all worry
that involves the environment
maybe due to lack of education
that the whole world is sick

Vivemos em um mundo complexo
diante de tantas tragédias
talvez seja algo com nexo
trabalhar educação ambiental nas escolas periféricas
We live in a complex world
in the face of so many tragedies
maybe it's something with nexus
working environmental education in marginalized schools

Precisa-se de uma grande revolução
no sentido de mudanças
pois somente a educação
nos permite ainda a esperança
A great revolution is needed
towards changes
because only education
still allow us to hope

a pandemia vem afetando a população
falta de atitudes e só lamentos
sabendo que toda destruição
são oriundas dessa "doença", a falta de conhecimento.
the pandemic has been affecting the population
lack of attitude and only regrets
knowing that all destruction
they come from this "disease", the lack of knowledge.

A educação de uma forma geral
precisa a aprender com a educação do campo
tendo foco na agroecologia e questão ambiental
resgatando saberes e assim cultivando
Education in general
needs to learn from rural education
focusing on agroecology and environmental issues
rescuing knowledges and thus cultivating them

Agroecologia tem sido pauta na discussão
e um estudo a se aderir na prática
garantindo que a futura geração
não seja tão prejudicada
Agroecology has been the subject of discussion
and a study to be adhered to in practice
ensuring that the future generation
don't be so harmed

sQuando me expresso dessa maneira
os acontecidos é preocupante
uma grande falha da educação brasileira
é colocar os problemas como uma "foto" de estante
when I express myself in this way
what happened is worrying
a major failure of Brazilian education
is to put the problems as a "photo" of a shelf

Então, a responsabilidade está em nossas mãos
Podemos ser atores/as de uma nova história
Se preocupar com a nação
É o que temos a fazer durante esta trajetória
So the responsibility is in our hands
We can be actors of a new story
We have to worry about the nation
That"s what we have to do during this trajectory

9. "From the Stars in the Sky to the Fish in the Sea" [*]

KAI ORCA

Living and working on Treaty Six Territory and the Homeland of the Métis, which includes the traditional territories of the Cree, Dakota, Dene, Nakota, and Saulteaux

Decolonized and sustainable lifeways require first and foremost that we establish sustainable relationships with our own bodies. In this chapter, I offer my own experience of learning from my children how to live authentically in my own flesh, embodied and relational experience which led me to question current mainstream social categories. Then, I evaluate how transgender literature written for children and youth by Kai Cheng Thom (2019), Vivek Shraya (2016), Gwen Benaway (2016), and Ivan Coyote (2012), and Two-Spirit poetry written by Joshua Whitehead (2017), can be used to help young people question fixed categories and develop more fluid, relational and ecologically-rich thinking. I create and interpret five lesson plans to show that gender identity begins with a person's own unique embodied experience, prior to language, that one's felt sense of identity is then given language through culture in a mutual exchange of embodied and linguistic meaning, that culture's gender categories are constructive and healthy when they reflect and enhance nature's diversity, and that the Indigenous gender history of North America should be our primary starting point for respecting and understanding gender on this land.

This chapter displays an experiential background through creative memoir to explain the basis out of which an understanding of Two-Spirit and transgender experience can grow. Then it demonstrates how to apply

[*] Thom, K. C. (2019). *From the stars in the sky to the fish in the sea.* Vancouver, BC: Arsenal Pulp Press.

Nicole Seymour's (2013) ideas about transecology to sj Miller's (2016) Queer Literacy Framework in teaching Two-Spirit and transgender children's literature, to help children to open their understanding of categories. Using Two-Spirit and transgender stories as theory to show how genre and gender co-create, how acts of naming and shaping experiences through story help to define and give shape to categories, this work demonstrates that Two-Spirit and transgender perspectives are critical for a renewed understanding of categorical thinking. Teaching children to have a more fluid approach to gender categories can help them experience the overlapping, interacting, and interconnected nature of all life on earth. This chapter demonstrates a methodology for using Two-Spirit and transgender literature to help children and youth question the basis of their embodied experiences and understand how these experiences are shaped through language, leading to a more fluid and relational practice of thinking. Fluid and relational thinking are critical for understanding the complex ecological interrelationships and messy processes of growth and evolution among life forms, as well as for navigating social interrelationships with all their chaotic and deeply emotional challenges around power and identity.

While the human propensity for categorization is useful for every day functioning in the world, it can also become destructive if the categories imposed onto the world do not actually fit what is there, and when these categories serve to maintain coercive hierarchies. How children and youth perceive the world depends both on their experiences and on the stories and labels they use to name those experiences. Understanding that categories are never final, but always open-ended and subject to change will help students to develop a more fluid and relational practice of thinking. This chapter leads to the conclusion that an appreciation for gender diversity goes hand in hand with appreciation for nature's diversity.

Nicole Seymour (2013) and Joan Roughgarden (2009) explore relationships between gender and ecology, theorizing a transecology, a method that points to the ways in which cultural treatments of trans bodies reflect cultural appropriations of nature and the ways in which the subjective experiences of trans people lead to subversive strategies to counter that appropriation (Seymour, 2013; Roughgarden, 2009). In this essay I apply concepts from transecology, that sexual and gender diversity are part of nature, and that a recognition of this diversity leads to more authentic understandings and encounters with nature, to sj Miller's (2016) Queer Literacy Framework, a set of ten principles for helping teachers develop educational lesson content which supports Two-Spirit and transgender students. I argue and demonstrate that Two-Spirit and transgender literature can help children to deepen their ecological awareness and strengthen their sense of agency in complex environments. While sj Miller's (2016) educational program focuses on supporting gender diverse students and showing how attention to gender diversity benefits all students, I argue that the knowledge and insights flowing from gender diverse people can help all students develop the ability to deal with complex cultural and ecological challenges, that these benefits extend beyond gender into all aspects of our ecocultural lives because they support relational as opposed to hierarchical thinking.

In his collection, *Teaching, Affirming, and Recognizing Trans and Gender Creative Youth: A Queer Literacy Framework*, sj Miller (2016) makes it clear that gender diverse students are in crisis, that teachers still fail to include and are afraid of including materials that would support these youth, such as rich information about gender diverse histories and literatures. Not only that, but the policies of most school districts in North America actively stigmatize gender diverse youth. Because gender diversity is mistakenly viewed as sexual activity, teachers are barred from teaching gender diversity until grade 7, 8, or 9. Any supportive material that does make it through this firewall meets the deadly "opt-out" requirement, whereby teachers are required to warn parents of a coming lesson that deals with gender diversity, giving them ample time to remove their kids and vacate the premises. Given this dire situation, in which the identities of some children are stigmatized in order to fabricate the supposed normality of the identities of other children, and its outcomes in the form of suicidality and trauma, sj Miller (2016) created a structure of principles to support teachers, a Queer Literacy Framework:

1. Refrains from possible presumptions that students ascribe to a gender.
2. Understands gender as a construct which has and continues to be impacted by intersecting factors (e.g., social historical, material, cultural, economic, religious)
3. Recognizes that masculinity and femininity constructs are assigned to gender norms and are situationally performed
4. Understands gender as flexible
5. Opens up spaces for students to self-define with chosen (a) genders, (a) pronouns, or names
6. Engages in ongoing critique of how gender norms are reinforced in literature, media, technology, art, history, science, math, etc.
7. Understands how Neoliberal principles reinforce and sustain compulsory heterosexism, which secures homophobia; and how gendering secures bullying and transphobia

8. Understands that (a) gender intersects with other identities (e.g., sexual orientation, culture, language, age, religion, social class, body type, accent, height, ability, disability, and national origin) that inform students' beliefs and thereby actions
9. Advocates for equity across all categories of (a) gender performances.
10. Believes that students who identify on a continuum of gender identities deserve to learn in environments free of bullying and harassment (Miller, 2016, p. 36)

I take Miller's (2016) Queer Literacy Framework as my foundation for educational work, but I would like to add the perspective that gender normative students benefit by learning from gender diverse people. Gender diverse students are certainly in need of social and education support, but they also have rich resources of intelligence and perspective that society needs in the face of our ecological crisis. I see gender inclusive education as that which makes it possible for this critical knowledge to be shared.

"Two-Spirit" is a term which was adopted in Winnipeg in 1990 to represent diverse linguistic terms in multiple North American Indigenous languages which create space for gender diversity (Roscoe, 1998). Until the oppressive practices of invading European-American and European-Canadian settlers forced two-spirit practices underground, Two-Spirit people were widely revered for having special and enhanced forms of knowledge and power (Roscoe, 1998): "According to Albert McLeod, the donor of the [University of Winnipeg two-spirit] collection, the term 'two-spirited' has its origins in a gathering of North American LGBT Indigenous persons in Winnipeg in 1990. McLeod (2013) defines the term as having 'the ability to reflect the male and female energies (genders and sexes) and forces that create life (e.g., humans, animals and plants) and that diversity within this realm is considered sacred and a component of the natural order (meant to be)'" (McLeod, in Lougheed, 2016). By taking on roles of the supposed "other sex" or by combining roles, a Two-Spirit person could express experiences that in mainstream North American culture might be registered as gay or transgender, but in Indigenous cultures mean having a special depth of skills and special access to the spiritual realm which generates respect and veneration. Instead of being stigmatized, as in European-heritage cultures, Two-Spirit people were, before colonization, thought to bring wealth and success, stability and balance to society. Stigmatizing European viewpoints toward gender diversity and sexuality led invading missionaries and soldiers to crush Two-Spirit people and traditions with excessive violence and vengeance, and many Indigenous people slowly came to adopt some of the attitudes of their colonizers and oppress their own Two-Spirit people (Roscoe, 1988). When I began to study Two-Spirit history, it changed my relationship with the North American land I live on. I began to feel that the land itself still carries the memories of thousands of years of respect for gender diversity. The devastating and brutal period of colonization still cannot outweigh the spiritual power of this history, which supports me every time I take a step. Currently, Indigenous people are starting to reclaim their Two-Spirit traditions and call their Two-Spirit people back into community, though the term "Two-Spirit" is not universally-chosen to express queer gender and sexual diversity among Indigenous people. Some communities prefer their own unique terms and understandings. The editors of *Sovereign Erotics: A Collection of Two-Spirit Literature* suggest that we add the term "queer Indigenous" to Two-Spirit to reflect the diversity of Indigenous queer experience (Justice et al., 2011, p. 6).

The term "transgender," has largely been attributed to Virginia Prince, who used it during the 1970's to distinguish herself from transexuals who pursued body modification, surgeries and hormones, but K.J. Rawson (2015), in research "conducted with Cristan Williams, ... found an earlier use of the term: psychiatrist John F Olivan used transgenderism in the medical text Sexual Hygiene and Pathology, published in 1965 ... to indicate an 'urge for gender ('sex') change'" (Stryker & Whittle, 2019, pp. 3–4; Rawson, 2015). Cultural support for this urge is documented through archeological finds by researchers who are able to step out of the binary lens which colors their vision (Ghisleni et al., 2016). The term "trans" connected with a gender diverse person appears in Roman literature in reference to Sabina, Emperor Nero's wife. She was one of the revered Galli, sacred women who, though assigned male at birth, went through a ritual involving removal of the testes in order to become acolytes of the Goddess Cybele in a tradition from Phrygia. She was described as "transfiguring" her body (Morgan & Power, 2021; Martin & Walsh, 2021). The term transgender "took on its current meaning in 1992 after appearing in the title of a small but influential pamphlet by Leslie Feinberg, *Transgender Liberation: A Movement Whose Time has Come,*" referring to people who do not identify with the gender which was assigned to them at birth based on their perceived reproductive capacities, regardless of how they choose to express their identities (Stryker & Whittle, 2019, pp. 3–4). In contrast to "transgender," "cisgender" has come to refer to people who do identify with the gender that was assigned to them at birth, leading to a sometimes problematic binary, since gender diversity actually applies to all people: "'gender,' as it is lived, embodied, experienced, performed, and encountered, is more complex and varied than can be accounted for by the currently dominant binary sex/gender ideology of Eurocentric modernity" (Stryker & Whittle, 2019, p. 3). Everyone can experience pressures to conform to cultural stereotypes associated with their assigned genders as uncomfortable and oppressive, or alternatively, affirming, at different times in their lives. I define Two-Spirit and transgender literature to be literature written by Two-Spirit and transgender people. Cisgender writers can create representations of Two-Spirit and transgender people, which may be powerful and insightful, but they are still representations, and not informed by the personal life experiences of Two-Spirit and transgender people. Other scholars take a broader position on Two-Spirit and transgender literature and consider it to be literature which deals with and represents these identities (Carroll, 2018).

Sj Miller's (2016) Queer Literacy Framework encompasses critical practices, which teachers need to understand, if they want to support gender diverse students, such as not assuming gender, letting students self-identify, realizing that gender is flexible, that while gender identity is personal and deeply-felt, gender itself is also socially constructed and reinforced through cultural media, that gender expression is a form of free expression which deserves protection, and that gender intersects culturally with a variety of other identity categories (Miller, 2016, p. 36). These principles help to create space for respect in highly-charged and too often transphobic school landscapes. Biologists like Joan Roughgarden (2009) and cultural ecologists like Nicole Seymour (2013) help us understand how cultural gender stereotypes and assumptions seep into and compromise the practice of science, and how cultural understandings of sex and gender impact our understandings of our world. Though sex is commonly understood as a biological phenomenon and gender as a cultural one, in reality the way we think of sex is highly colored by the social construction of gender. We cannot really understand and have an authentic relationship with our natural surroundings until we understand how gender constructs impact this relationship.

Environmental education (EE) involves encouraging children to develop healthy relationships with their surroundings, but adults often forget that the very first relationship a child works to develop is with their own body, in that primal space and time when a child first realizes that they have agency to move arms and legs, to roll and crawl, to engage in a muscular and sensual exploration of a world full of energy, color and sound. In order to develop personal and physical integrity, immediate physical sensations need to be connected with cultural information. Yet while cisgender children are usually able to make the connections between body, self, and society, Two-Spirit and transgender children in North America eventually discover that their embodied sense of self, often expressed as early as at two years of age, goes unsupported by their human environment: "Ultimately, it is not just transgender phenomena per se that are of interest, but rather the manner in which these phenomena reveal the operations of systems and institutions that simultaneously produce various possibilities of viable personhood, and eliminate others" (Stryker & Whittle, 2019, p. 3). Two-Spirit and transgender children too often experience themselves erased and written over, even though most North American Indigenous communities had means and ways of supporting their Two-Spirit children, which honored the powerful spirits of these children, before the historically recent arrival of Europeans (Roscoe, 1998).

My children taught me to listen to my own experiences, when I had learned to pave them over with the heavy concrete of categories which were not my own. Perhaps the essence of EE involves listening to the children, so that we can listen to our own bodies and the earth in fresh ways. In teaching children to express their own gender identities more authentically, we can also teach them to respond more authentically to the world around them, and by connecting with the earth more openly, they can be more at home in their gendered or ungendered bodies. Gender education is deeply connected with EE because it is only through our experiences of our own bodies in tune with our surroundings that we come to know the natural world.

Our Story

My child Sky had a very hard time learning to recognize the difference between left and right. Even as they grew older Sky still seemed confused. My partner and I grew increasingly frustrated. What was wrong, we thought, with a child who could not grasp this very essential and basic difference? One day, when I as usual was being a bit impatient with Sky over this problem, Sky asked,

"Why are you always changing it around?"
"What do you mean, changing it around?" I asked.
"First left is over here," Sky gestured left, "and right is over there. But next time you change it all around and left is over there," Sky gestured to the right, "and right is over here."

Suddenly I understood. We had forgotten to coach Sky in one very important point, that left and right only make sense in relation to your own body. This astounded and humbled me. It reminded me how often we fail to remember all the steps that go into learning something. It also speaks to the importance of embodied learning. We had failed, as educators and parents, to help

Sky connect their knowledge of their own body to their knowledge of the environment.

I learned this lesson again when my daughter Elsa was two years old. We were walking down the sidewalk in Boulder, Colorado. The sun was shining. The rust-red Flatirons rose upwards to the west, touching a blue spring sky, while the snow on North Arapaho peak glittered high on the divide. The earth was fresh with the smells of young plants growing and bright with the sight of new buds opening on tree branches, exposing tender pink flowers. As my daughter and I chatted, I grew confused because she seemed confused. She had been assigned male at birth. Dutifully going along with my society's gender system, I responded to her manner of self-identifying herself as a girl with,

"No honey, you are a boy."

Her scream reverberated against the high altitude air, and made the whole world seem dark and hollow. It was a scream of terror, of a small child discovering that she was invisible, that her knowledge of self was ignored and steamrolled over by forces that she did not understand. Just as I failed to listen to Sky and failed to help Sky connect body, self, culture and world, I failed with Elsa as well. I discounted her own self-knowledge and over-wrote it with the dictates of a colonial gender system that had already destroyed the lives of countless gender-diverse people. I became the conduit for cultural forces of destruction and complicit in the erasure of my own child.

By the time I understood what I was doing, six years had passed, and my eight-year-old daughter had suffered enormous psychological stress. My daughter is now fifteen. In the past seven years we have experienced society's rejection of transgender children firsthand. The supposedly progressive schools and churches, which we have tried to be a part of, have refused to teach and talk about transgender children, have attached shame and secrecy to my daughter, have sexualized and fetishized her child's body. I have been told that teaching children about gender identity and expression will confuse and disturb other children, that it is "age inappropriate."

What kind of society is so uptight about its own gender system that it refuses to discuss the way it conditions children into stereotypical and damaging behaviors? Just how arbitrary and unnatural must such a gender system be if it cannot bear any critical discussion, if it is willing to sacrifice and destroy innocent children rather than open itself to the reality of gender diversity? My children have attended eleven different schools in four different states and provinces, Colorado, British Columbia, Idaho, and Saskatchewan, and none of those institutions were able to fully include my children in a way that could support them and build their pride and integrity.

At Sunrise Waldorf School in British Columbia my daughter was bullied every day, and her teacher told me that I was to blame for that bullying for letting her dress in a way that expressed her sense of self. She was forced to use the "boys" washroom, which was decorated with snakes and lizards that glared at her from the walls, in her mind screaming that she did not belong. She avoided that washroom and began to develop a serious intestinal disorder, which caused her to lose control of her bowels. Children looked askance at her, reminding her that they considered her strange. At Syringa Mountain School in Idaho, administrators and teachers pointed my daughter out to others from a distance, watched her and whispered, teaching her that even the adults marginalized her. The school community was uncomfortable with us and ultimately rejected us completely. At Creekside School in Colorado my daughter's teacher told us that they do not teach about transgender identity in second grade because it is believed to have something to do with sexuality, rendering my daughter invisible, yet sexualized and shamed. At Eisenhower Elementary School in Colorado the teachers who promised me that they would include transgender stories and histories in the curriculum, to help my daughter develop pride and foster inclusion, turned out, after taking much of my time and energy engaging on this task, to have told the principal at the start of the year, that they would do no such thing. At Community Montessori School in Colorado the principal was prevented from bringing Phoenix: Colorado's Transgender Community Choir to perform my story *Raven's True Self* without the damaging opt-out option, by Boulder Valley School District, a district which prides itself on being trans-inclusive. The Unitarian Universalist Church of Boulder's Educational Program refused to teach about gender identity, leading to the marginalization of my children, and ultimately a physical assault on the church grounds. At Brunskill Elementary School in Saskatoon my children were routinely accosted by other children with taunts about their perceived gender identities and sexualities. Except for one hour of programming put on by OUTSaskatoon, preceded by the mandatory opt-out option which attaches a sense of shame to the material, there was no explicit curriculum in the classroom about cultures and histories of gender diversity. These forms of exclusion lead to the high rates of harassment and violence that gender-diverse children and youth experience, which help to prepare them for the high rates of homelessness and joblessness that they can expect to experience as adults.

We live in a binary system of gender segregation, an apartheid system, based on a distorted interpretation of biology, that insists on assigning gender and indoctrinating children to fit assigned gender identities and roles, as psychologist and philosopher of science Cordelia Fine has demonstrated in her phenomenal psychological study *Delusions of Gender: How Our Minds, Society, and Neurosexism Create Difference*. We assign meaning to bodies at birth in a binary way. We segregate the humans whose bodies and selves we have defined. We shame, taunt, threaten, beat and batter children into expected dress styles, mannerisms, behaviors, activities, roles, and jobs. If this system were biologically natural, as so many people claim, then why do we need to submerge the children in an ocean of endless media stereotypes, combined with parental and societal punishments and rewards, in order to achieve the outcome of gender conformity? According to Fine, parents and educators are complicit in shaping malleable young brains and bodies without permission. Parents treat their children differently according to gender assignment, creating the very differences which they believe to be biological (Fine, 2010, p. 198). Furthermore, the same societies which insist rigidly on assigning gender based on assumed material physical natures also insist on bending and shaping the earth according to human design. This oppressive cisnormative system creates in gender-diverse people an experience of questioning, like the experience of Frantz Fanon [who] closes his exploration in *Black Skin, White Masks* (1967) with a prayer, " 'Oh, my body, makes me always someone who questions' " (Mignolo, 2011, p. 274). Historical and anthropological evidence demonstrates that most Indigenous North American societies prior to European invasion responded to children expressing gender diversity with respect and accorded them honored roles (Roscoe, 1998). It is no surprise that these same Indigenous societies place great emphasis on healthy relationships, on listening to and respecting all beings, rather than trying to force them into preconceived roles and identities.

As I have learned to listen to my children, they have opened my eyes and heart to the enormous range and diversity of human and biological nature. Identities and relationships are complex, rich and multifaceted. As biologist and transwoman Joan Roughgarden (2009) points out in *Evolution's Rainbow: Diversity, Gender, and Sexuality in Nature and People*, the only actual tendency toward a sex binary in nature is that of gamete size. In most species that reproduce sexually, there is a divergence in gamete size, with "a near-universal binary between very small (sperm) and large (egg), so that male and female can be defined biologically as the production of small and large gametes, respectively" (Roughgarden, 2009, p. 26). Everything else in nature and humans is a rainbow, from sexual practices to family structures to parenting strategies to stability of sex to gender expression (Roughgarden, 2009). Gender and sexual diversity are an integral part of nature's critical range, nurturing the complexity of cooperative relationships that sustain life. Our current failure to recognize this and our determination to roll over both children and earth with our preconceived ideas about how things should be is a shame and a disaster, creating gender-traumatized humans and a severely endangered and depleted earth. If we are going to develop sustainable thinking and sustainable education, we must be able and willing to crack open the gender system that our mainstream North American culture currently lives by, and which feeds into the unnatural idea that the size of a person's gametes will determine their feelings, thoughts, and skills.

We must begin to allow children to express the full range of their humanity. Imagine a world where every human is allowed from the earliest moments to cry, to feel, to care, to listen to animals, to watch flowers open and feel the exquisite beauty of it all, to grow their hair long and feel it blowing in the wind, or cut it short and feel that delicious lightness, to dance and feel their body moving to a rhythm, to be strong or delicate, to love colors and design, or embrace simplicity, to be dominant and submissive, to fly airplanes or rocket ships, or tend to a child and feel that sense of responsibility, love and care. Imagine a society that is open and brave enough to examine itself and realize that it does not need to destroy the lives of some children in order to support the identities of others. Every human has the right to grow into and establish their gender identity, express the full range of their gender experience, and develop their own love relationships freely, without shame, just as they have the right to know the earth and its ways more authentically.

As I watch my daughter Elle growing into a young woman and my child Sky developing their own genderqueer identity, as I feel my own lifelong genderqueer nature flowing through me, I am struck, as well, by the connections between gender categories and borders of all kinds. In trying to establish certitude and solidity we create borders. Borders separate one thing from another. Increasingly a "proliferation of segregating, militarized borders around the world" are "situated on the boundaries of inequality" (Miller, 2019, pp. 56–7). As people struggle to protect themselves from the "catastrophic results of elite injustice, corporate lies, and collective thoughtlessness," and from their own willful self-serving ignorance and unwillingness to take constructive action, they build borders which are as doomed to fail as sea walls designed to keep out the powerful surges of a rising ocean (Miller, 2019, p. 56). As Thomas King (1993) demonstrates through his short story *Borders*, we are always trying to establish boundaries and borders. They help us to clarify and distinguish, to feel that we understand things, but borders are never solid. For many years I have moved back and forth across the Canadian-U.S. border. I used to think that I had to take a side. I had to be either American or Canadian. I finally came to realize, with the help of Thomas King (1993), that it simply is not my border. My attachments and experiences bridge that particular line. My world is bigger and inclusive of events, people, and histories on both sides, and while that might make for a more complex identity, ultimately it is richer and more nuanced than my identity would have been if it had encompassed only one side, or if I thought I had to take a side and reduce and disparage everything on the other side in order to expand and glorify everything on my side.

The same is true for my gender identity. When my daughter was finally getting through to us and claiming her identity, all the while connecting with Elsa of the movie *Frozen*, who also had a terrible secret that seemed to threaten society, she told me, "I felt like I had a terrible secret." I recognized that as a familiar feeling. All my life I never really identified with my gender assignment, but until my children taught me otherwise, I thought I had to go along with it. Ultimately my truth is that, just as the U.S.-Canadian border is not my border, the border between people who experience themselves as men and those who experience themselves as women is also not my border. My experiences encompass both genders. I am a genderqueer North American.

Just as we need to look more carefully and critically at our gender system, we need to examine borders of all kinds. Our tendency to create borders and boundaries between things in order to make sense of those things needs to be constantly reexamined. Borders are never solid, never fixed for all time. Borders prevent us from remaining open to what is there, rather than assuming that we already know and understand it. Whether it is the gender identities of our children, the nature of our national identities, or the rich capacities and fragilities of our earth, we need to learn to listen.

As I walk along Boulder Creek, reflecting on this chapter, I am looking at a brilliant blue Boulder, Colorado sky, breathing in the warm smell of pine, which always reminds me of summers with my grandparents on Evergreen Lake in the Colorado Rockies, waiting for a grasshopper to make that startling leap, while thinking of the summer clouds swirling over Saskatoon and over the South Saskatchewan River as it makes its great way across the Prairies, feeling the strength of my masculine-feminine body and knowing that my children will have the resilience that comes from crossing borders. Instead of erasing children, let us erase the borders that divide us. As I take in this combination of sense, sight, sound, and memory, I am thinking of four important aspects of sustainability that link environmental and gender education, and the ways in which specific works of Two-Spirit and transgender literature can help teachers and students to make these links:

(1) Teach awareness of one's own embodied experiences.
(2) Teach recognition of cultural gender stereotypes.
(3) Connect gender diversity with natural diversity.
(4) Teach awareness of Indigenous gender systems.

Connecting Environmental Education with Gender Education

A variety of research confirms the problems which I and my children experienced personally, that teachers are still largely uncomfortable addressing Two-Spirit and transgender topics, that Two-Spirit and transgender students remain unsupported and at risk, that an institutionalized silence hangs over these issues and lives like a pall (Miller, 2016; Brant, 2016; Shelton & Lester, 2016; Ehrensaft, 2014; Travers, 2014; Meyer, 2014; Grace, 2013). These realities make it all the more imperative that we open up spaces for gender-inclusive education and help teachers, parents, and students alike to understand why such education is linked with health and sustainability. As Nicole Seymour (2013) maintains, "new models of gender and sexuality emerge not just out of shifts in areas such as economics and medicine, but out of shifts in ecological consciousness" (p. 30). Our ability to locate Two-Spirit and transgender experience within the fabric of a healthy and supportive organic framework is directly linked with our ability to understand the actual interconnected ecological complexity of the world around us.

Sj Miller's (2016) edited collection, *Teaching, Affirming, and Recognizing Trans and Gender Creative Youth*, includes not only his Queer Literacy Framework, ten principles to help educators design lessons and programming which can help gender diverse students, but also specific lessons designed by a variety of teachers and scholars for students in classrooms from kindergarten through high school applying these principles. These lessons offer helpful models for introducing gender diversity into K-12 classrooms and help students to question the gender norms which may not support equity and healthy identity formation. British Columbia's Sexual Orientation and Gender Identity (SOGI) policies (Government of British Columbia, n.d.) and curriculum guide likewise provide a framework for creating gender inclusive school spaces. Both of these sources, however, tend to focus on the task of supporting gender diverse students, with the supportive energy flowing from the surroundings into the gender diverse person, who is inevitably cast as a victim. My work reverses that flow and points to the healing and rich perspectives, the complex and flexible thinking, and the powerful spiritual history which literature created by gender diverse people can bring to all children and youth. As gender diverse children see people like them as sources of knowledge, they will do more than survive. They will thrive. Seymour's (2013) point that gender diversity is a powerful ecological lens can be added to Miller's (2016) principles to develop sustainable educational practices and show that gender diverse people have historically served the role of helping their communities reach beyond limiting frames and embrace healthy change.

We can make the connections between EE and gender education through the following practices, leading to enhanced learning outcomes in both areas. These include allowing for embodied awareness, understanding cultural gender stereotypes, making connections between natural and cultural diversity, and teaching about Indigenous gender systems. I demonstrate these practices with reference to five lesson plans available on my *Earth Tide* website on Patreon (<https://www.patreon.com/EarthTide>). In this chapter I evaluate and interpret these lesson plans, using a transgender-ecological lens, to synthesize my central points, that gender identity begins with a person's own unique embodied experience, prior to language, that one's felt sense of identity is then given language through culture in a mutual exchange of embodied and linguistic meaning, that culture's gender categories are constructive and healthy when they reflect and enhance nature's diversity, and that the Indigenous gender history of North America should be our primary starting point for respecting and understanding gender on this land, not frameworks imported from Europe which reject natural and cultural diversity. Most importantly, the critical perspectives of gender diverse people have and do serve to enrich our understanding of the world around us. These proposed lesson activities relate to Saskatchewan English language learning outcomes supporting a healthy development of identity, sense of community, and responsibility, health outcomes relating to awareness of one's own feelings and needs, a healthy sense of self, positive relationships with others, and social studies outcomes dealing with cross-cultural respect and Indigenous cultural understanding, as well as cross-cultural understanding of the meaning of spiritual awareness and inner self. Relevant Saskatchewan science learning outcomes involve the understanding that classification systems are always open to change, that "scientists develop classification systems to meet their needs at a specific time and modify classification systems in light of new evidence. Given this, science teachers can help students see connections between modifications to the classification of plants and animals over the past few centuries and changes to our understanding of gender diversity among humans" (Saskatchewan Ministry of Education, 2015, p. 54). Students will learn that human diversity is part of natural diversity and that human understandings of both are interrelated.

Five Lesson Plans

Full Lesson Plans Available At: <https://www.patreon.com/EarthTide>

(1) Grade One: Teach awareness of one's actual feelings, strengths, abilities, and desires connected with personal sources of knowing, using Ivan E. Coyote's (2012) short autobiographical story *No Bikini* (Identity)

- Students draw and discuss two pictures, one to represent the kind of freedom Coyote describes to express oneself freely and feel alive in the world, and one to represent a situation in which people around one expected one to be or act in a way that made one uncomfortable. They reflect on the sources of social expectations and consider means of dealing with them which allow one to still be oneself. The very sensual nature of Coyote's experience helps the children understand gender identity to be rooted in deeply embodied physical experiences.

(2) Grade Two: Identify and question cultural gender stereotypes and understand cultural change with Vivek Shraya's (2016) short story *The Boy and the Bindi* (Community and Culture)

- Students draw, discuss, and perform to represent the boy's bindi, what it means to him, how it gives him a sense of harmony, balance, and connection, and what the cultural expectations are surrounding its use. They consider how social change can happen when one child expresses his needs and encourages his mother to stretch cultural traditions to meet those needs, demonstrating how strong human relationships can expand the parameters of social acceptance.

(3) Grade Three: Connect gender diversity in humans with gender diversity in other species, and with issues of diversity in nature and society, using *from the stars in the sky to the fish in the sea*, a short story by Kai Cheng Thom (2019) (Environmental and Social Responsibility)

- Students notice and draw forms in nature, think about the relationship between form and function, and reflect on how change over time may have influenced the related shapes of different organic forms. Then the students write, draw and discuss Miu Lan's four different ways of showing up at school. From completely open, to prickly and defensive, to conforming, to realizing that in being themselves they can help the other children become more open and free, Miu Lan's different approaches to school represent different strategies available to a gender creative person in dealing with a gender normative environment.

(4) Grade Four: Teach about Indigenous gender systems which involve respect for gender diverse individuals. Look at the issue of appropriation, how settler Canadians are claiming to have connections with these forms of knowing, with Gwen Benaway's (2016) poem *Lake Michigan*. How should we handle these issues of cultural appropriation, and is it appropriation when someone like Gwen Benaway feels a sense of connection to Indigenous life through her family history, but is not necessarily claimed by current-day Indigenous communities? (Indigenous Ecological, Social, and Spiritual Knowledge; Cultural Appropriation)

- Students work together first to create a large painting of Lake Michigan and add features to represent the way land holds history, connections between water and human bodies, the undercurrent of the lake as ancestral lodge, and the contrast between colonial quantified ways of defining people and more intuitive and relational ways of knowing, and second to discuss issues of cultural appropriation, the Two-Spirit identity claims of the poet, the controversy surrounding these claims, and to write a paragraph about these issues.

(5) Grade Five: Teach how Joshua Whitehead (2017) creates an image of an emerging two-spirit powwow dancer to imagine the energy of Two-Spirit people reclaiming their place in their communities and on the land. (Indigenous and Two-Spirit Cultural Resurgence)

- Students respond to Whitehead's poem THEGARBAGEEATER by creating a representation of Whitehead's powerful viral cyborg powwow dancer emerging from the apocalyptic Canadian oil wastelands, a combination of virus and machine, a "steeltown ndn moloch / supersonic thunderbird / graveyard scrapyard cyborg" (35). In addition, they write a paragraph to explain their model.

The Importance of Story and Art as Educational Method

Story allows children to enter into the experiences of another, to live through shared experiences, and puzzle over the complexities and challenges of social life and personal identity, like Ivan Coyote (2012). Building lesson plans out of literary art, which require the children to create art in return, expands the child's ability to reflect and source personal knowledge. In lesson one, children read and respond to Coyote's (2012) autobiographical short story *No Bikini* and draw two pictures, one to represent one's own feelings about one's physical experience, and one to represent how one is seen by others (inside and outside views), after listening to and discussing the story:

> ➤ Discuss the way the phrases and sentence "six weeks of bliss," "diving boards and cannonballs and backstrokes" and "water running over my shoulders and back felt simple, and natural, and good" express joy, power, and freedom.
> ➤ Discuss the way the phrase "jump into the deep end" expresses the feeling of overcoming fear and exploring a big unknown place. Ask them if they have jumped into the deep end of a swimming pool and what that felt like. What swimming skills did they need to master before they felt safe to do so?
> ➤ Discuss other situations in which they felt free and full of joy after overcoming fears.
> ➤ Draw a picture of a situation in which they also felt this kind of freedom or overcame fears to enter a large and unknown place or space, with a focus on their sense of self and identity in this moment. Ask them to write one word on the picture to describe their picture, or at least one letter, which represents that word.
> ➤ Discuss the way the phrases and sentence "feel like a crime," "be ashamed," "eyes straight ahead on the road," and "Her eyes were narrow and hard, and trying to catch mine in the rearview mirror" express the way Ivan's community and family expect Ivan to be a certain way and fit into a particular group or category, namely the group of girls.
> ➤ Ask them if they have ever felt like they had to be or act like something that they are not.
> ➤ Ask the students to draw a picture of a situation in which people around them wanted them to be or act in a way that made them uncomfortable. Have the students depict how they think people were seeing them in this moment, along with their feelings in response. Ask them to write one word on the picture to describe their picture, or at least one letter, which represents that word.
> ➤ Have the students share and explain their pictures with a learning partner.
> ➤ Have students discuss the difference between their two pictures, first with a learning partner and then with the class, in regard to the following questions:

- Why do people around us sometimes want us to be something, to fit into a group, or act in a way that does not fit with who we are?
- Can we know who we are from the way we feel inside our bodies?
- What strategies can we use when we experience a conflict between the way we know ourselves to be and the way people expect us to behave?
- What strategies did Ivan use in this situation?
- Were these strategies effective? Why or why not?
- Did Ivan grow up to be a woman?

Because story involves multiple perspectives, characters in relationship working out differences and sustaining the possibility of care across divides, using story to create lesson plans enhances the child's ability to deal with complex social challenges, as in lesson two:

➤ Read and respond to Vivek Shraya's (2016) story *The Boy and the Bindi* by creating a map of gender stereotypes operating in the story and contrast those with the child protagonist's own personal source of strength.

➤ How does the child both draw on and challenge cultural assumptions?

➤ What role does the mother play in helping the child find a place for himself within cultural traditions?

➤ Compare this to your own life. In what ways do you draw strength from your community's values? In what ways do you challenge them?

➤ Help the students notice the five-beat rhythm. Have them clap the rhythm while you read parts of the story. Draw their attention to the paired rhyme words.

➤ Discuss the following questions:
   - Why do you think Vivek Shraya chose to give the story this rhythm and these rhymes. How does the story feel different when it has these patterns? The rhythm is based on five. What things do you know that come in fives? The paired rhyme words draw attention to pairs. What things in the story appear in relationship with one other thing?

➤ Explain that a red bindi is traditionally worn by married women in India, especially Hindu women. Even though they are usually worn by women, the mother does not hesitate to give one to her son.

➤ Discuss the following questions:
   - What is a bindi? Why does the child want one? Why is the mother's bindi red and the child's bindi yellow?

- How does the bindi connect the mother with her own mother?
- If this is traditionally something shared between women, why does the mother agree to give one to her son? What might she have done instead? How would the boy have felt if she had refused to give him a bindi and told him that they were just for girls?
➤ Explain that Hinduism is a religion in which there are many manifestations of creative power, many forms of God, and a strong focus on achieving an internal sense of harmony.
- What does the bindi do for the child?

The open-ended creativity of story and art, the way multiple possibilities interconnect and lead to new insights, parallels evolutionary creativity in nature, the way organic and inorganic forms take on shape, color, range of function and interrelationship with other forms. Exposing children to the artfulness of nature and story inspires joy and creativity. In lesson three, the children learn that creativity in humans is part of the larger range of creativity in nature, that humans develop identity by connecting with natural forms and processes:

- Are fish fins like feathers? How is it different to move through air, compared to moving through water, compared to walking or crawling on land? What kinds of bodies are best for moving in these different spaces?
- Are humans like bears, or wolves, or dogs or cats, or monkeys? How and how not?
- What are tails for?
- How are different parts of nature related to other parts of nature?
➤ Explain the difference between form and function, that form is the shape of something, and function is what it can do. Ask the children to think about the way the form of something helps it do something. Leaves are often broad and flat in order to catch sunlight, just as ears are broad in order to catch sound. Bark protects trees, just as skin protects animals. Bodies are designed so that they are useful in the place where each animal lives.

Miu Lan's first approach to school is to express their joy and excitement by growing a peacock tail and a coat of tiger stripes.

- Why do the school children not want to play with Miu Lan?
- Why does the boy ask Miu Lan what they are supposed to be, and then pull their tail?

Miu Lan's second approach to school is to wear a turtle shell and porcupine spines.

- What is Miu Lan hoping to accomplish by arriving at school this way?
- Why do the school children point and whisper?
- Does this help Miu Lan get along at school?

Miu Lan's third approach to school is to dress the way the boys dress.

- Why do the boys accept Miu Lan for a game of baseball?
- Why do the children object when Miu Lan begins to play hopscotch with the girls?
- What is Miu Lan's response?
  - ➤ Look closely at the picture of Miu Lan in their mother's arms being comforted.
- Where are all those tears going?
- Can you make an ocean of tears?
- How sad is Miu Lan?
- What does Miu Lan's mother explain to Miu Lan?
  - ➤ What is different for Miu Lan when they go to school the fourth and last time in the story?
- Why do the children respond differently this time?
- What do the children learn from Miu Lan?

Poetry concentrates literary art with heightened attention to figurative language, rhythm, rhyme, sound, and structure. In lesson four, Gwen Benaway's (2016) poem *Lake Michigan* allows the children to experience the water of the lake as a retainer for historical memory and for the subconscious, while considering who gets to speak for Indigenous people:

> ➤ Have the students sit around a large panel of paper and work together to paint a large blue lake, while discussing the following questions, in relation to the poem's words: "where my ancestors / grew and died / along the shoreline / of every waterway" and "we began in you, / blue light holds us / in your sunken veins, / Lake Michigan." Be sure to write or have a student write all the phrases which you discuss from the poem onto the painting as you work together.
> - How is water a space that holds memory?
> - Why is Lake Michigan connected with ancestors?
> - Why does the water of the lake become a medium for entering into a sense of relationship with one's origin?

- How does the metaphor "your sunken veins" connect human bodies with the Lake's body?
- How does blue light make you feel?

➤ Have the students discuss the discrepancy between Gwen Benaway's own claim about her identity and the way some Anishinaabe and Metis communities contest that claim. Explain the harm done to Indigenous people by settler people claiming to speak for them.

- What happens in Canadian culture when a person feels a connection with Indigeneity, through family stories and claimed ancestry, but that person does not actually have connections with living Indigenous groups?
- Who gets to decide who is Indigenous? Why does this matter?

In lesson five, Joshua Whitehead's (2017) powerful image of a returning reemerging ancient and postmodern Two-Spirit powwow dancer allows the children to reflect on the way Indigenous people are connecting past and present, with resilience. By creating a model of this dancer, out of recycled materials and their own imagination, the children experience the moment of resurgence. The powwow dancer comes with the power to rearrange, reclaim and rewrite historic wrongs:

➤ Explain Two-Spirit people were traditionally highly respected in most North American Indigenous societies, for bridging and connecting between men and women, and between the community and outside forces. Their combination of masculine and feminine gave them special insights to work across numerous divides. They were healers and connectors, and associated with success, health, power, and wealth. Their skills especially involved helping people sustain healthy relationships with each other, with other beings, with natural forces, and with other communities. They were destroyed by the arriving Europeans who viewed anyone expressing gender diversity as somehow polluted and wrong. Alex Wilson's (2011) chapter on life experience and ideas in *Safe and Caring Schools for Two Spirit Youth* can guide this learning.

➤ Discuss the way Whitehead (2017) uses words on the page to create an image of a powerful cyborg powwow dancer emerging from the apocalyptic Canadian oil wastelands, a combination of human and machine, a "steeltown ndn moloch / supersonic thunderbird / graveyard scrapyard cyborg" (35). Notice the way Whitehead uses colons to mark out the powwow drumbeat, one one two two three four: ": : :: :: ::: :::::" Explain that Moloch was an ancient god associated with sacrifice from the area of the current-day Middle East, an area of ancient civilization and fertile agriculture. Explain that the Thunderbird is a powerful spirit in many North American Indigenous cultures who casts lightning and creates thunder by clapping their wings, has different manifestations in different cultures, but generally fights underwater spirits and is connected with the sun. Explain that a cyborg is short for "cybernetic organism" and is a machine creature, a creature that combines living and machine parts.

- What is the effect of combining the words moloch, thunderbird, and cyborg in one phrase? What qualities are combined by doing so?

All five lessons draw on story and art to engage students holistically with complex problems, using important themes and issues, image, metaphor, rhythm, sound, drawing, writing, modeling and movement to source embodied knowledge, identity gender stereotypes, connect natural and cultural diversity, and trace the gender diverse history of North America.

Themes and Issues

From these five works of Two-Spirit and transgender literature, children can discover the power of their own embodied knowledge, like that of Ivan Coyote (2012), to counter gender assumptions and gender assignments. They can learn to think more deeply about gender from different cultural perspectives with all five writers, developing a more critical understanding of how cultural categories inform and influence their experiences. They can think about cultural change, and how individuals, acting out of deeply-felt personal experiences and needs, and with the help of family and friends, can contest and shift cultural assumptions and traditions, to make them more inclusive and more compatible with human needs. They can discover with Kai Cheng Thom (2019) the wide range of interrelated forms and functions in nature, that seemingly distinct categories may in fact be connected through relationship with others, that species can shift over time, and that individuals can celebrate the way they experience nature's diversity in their own bodies and lives. They can also discover the rich gender diverse history of North America with Gwen Benaway (2016), begin to understand the violent and devastating history of the colonial era, and develop compassion and a commitment toward finding a way to heal. They can learn the importance of letting Indigenous people speak for themselves and avoiding Indigenous identity claims unsupported by actual membership in Indigenous communities. With Joshua Whitehead (2017) they can experience the powerful forces of Two-Spirit resurgence, the way Two-Spirit people are reclaiming respect and place in community. All five lessons counter the idea that we have to measure, define and control bodies, identities, possibilities, land and nature, and encourage respect for the emerging qualities, the uncontainable possibilities, and the unexpected relationships that emerge when we open our thinking, our feeling, and our experiences.

Teaching Fluid Thinking

Fluid thinking helps children understand that what they see on the surface is only a small part of what is really there. They have to probe more deeply to comprehend gender identity. They have to think about where their sense of self really comes from, whether it belongs to them or to people who might have cultural or personal investment in asking them to conform to roles and patterns. Like the child in Coyote's (2012) *No Bikini*, they can wonder why the parents are so concerned about trying to have the child identify as a girl and conform to social expectations. They can relive their own embodied feelings around these kinds of experiences and think more critically about social expectations. From Benaway's (2016) poem *Lake Michigan*, the children can learn that the common Canadian celebratory settler story of progress is just the surface view of a more troubling history of erasure.

Fluid thinking involves letting go of assumptions and being courageous enough to step into territory which might be new. Both the child in *No Bikini* and the child in *The Boy & the Bindi* challenge cultural expectations. Just as gender diverse people throughout human history have been revered for their ability to see from the outside of expectations, to help the more normative people, who were busy leading normative lives, find ways to shift their norms when conditions required adaptations for the community's health and survival, these children help the community by going against expectations, either with or without the help of parents, and helping people to shift their understandings. Coyote may have struggled with parents as a child, but I personally experienced them in performance in Whitehorse, Yukon, in the summer of 2018, exchanging stories from the stage with family members in the audience who had congregated specifically to celebrate Coyote's storymaking gifts. So clearly, Coyote's parents did change their understanding and expectations over time. In Shraya's (2016) fictional story, the mother has the wisdom to support her child with the gift of a bindi. Since the bindi holds such power as a cultural symbol, what she really tells her child is that there is a place for his feminine power too, that he has a place in the flow of culture. She empowers him, just as he empowers her.

Fluid thinking involves understanding that categories are open to change, that the multitudinous and diverse forms and patterns in nature, including human nature, overlap in order to make life more adaptable. In nature conditions change, and change requires a changed response. Forms allow specific functions, but functional needs can change. Being open like the child in Thom's (2019) story *from the stars in the sky to the fish in the sea* allows one to open one's heart to possible change and to celebrate the diversity of interconnected life. There are so many resemblances between different life forms because they are in fact related.

Teaching Relational Thinking

Relational thinking involves the knowledge that beings are interrelated, both within and between species. Within families people need each other. The protagonists and speakers in all five works of literature search for connection with family members and ancestors. Connection works across time and space. Relationships can cause tensions, which can provoke thought, understanding, and change. The parents in Coyote's (2012), Shraya's (2016), and Thom's (2019) stories are pushed to consider new forms of relationship. The speaker in Benaway's (2016) poem stands alone by Lake Michigan, calling her family and people back into relationship, yet Benaway's claim to Anishinaabe and Metis identity is contested by those communities. The Two-Spirit speaker in Whitehead's (2017) poem, Zoa, comes powerfully raging out of the wasteland, out of the garbage heap of history, of destroyed and castaway lives, like a viral force of nature. Two-Spirit and transgender bodies, which, since the European invasion into North America, have provoked violent fractures in relationship, force us to rethink communal capacities to include and respect differently-bodied people, and to rethink even normative forms of embodiment. Do we frame difference as threat, or as opportunity? Two-Spirit and transgender literature encourages special and heightened capacities to develop the kinds of relational thinking that are so central to Indigenous life. Instead of rejecting forms that are new, we can accept them as part of our world and work toward entering into relationship, work toward incorporating them constructively into the shared frame. Without appropriating Indigeneity, we can still accept the fact that North America is Indigenous land, and Indigenous ways of thinking should have flowed outward to influence new arrivals here. All North Americans can learn from and benefit from Indigenous ways of knowing.

Conclusion

Two-Spirit and transgender literature does not just serve to help educators create understanding for gender diverse students. Gender diverse perspectives are a powerful source of knowledge and openness. Historically, gender diverse people served to help their communities move beyond restrictive ways of thinking which limited their ability to survive changing conditions. Applying the insights of transecology, which demonstrate that Two-Spirit and transgender knowledge can provoke complex understandings of nature and culture, to queer inclusive educational practices, we can begin to discover the multiple ways in which Two-Spirit and transgender literature can expand fluid and relational thinking in children and youth, be a source of environmental knowledge, and a resource for sustainability. In this chapter, I have asked these important questions: Are we doing the foundational work of helping our children orient themselves in the world, connect with the sheer joy and physicality of their own bodies and discover relationships with the multifaceted forms and forces that surround them? Are we helping them understand how the cultural names, the lefts and the rights, are actually connected through their own bodies to the world? These names only make sense when the embodied connection is there. Most importantly, are we listening to what our children are telling us as they work to connect their own embodied experiences, through human culture and language, with the world? I have found means to answer these questions in literature by Two-Spirit and transgender authors and in frameworks for using this literature in crucial lesson plans to encourage embodied awareness, knowledge of cultural gender stereotypes, understanding of connections between natural and cultural diversity, and respect for Indigenous gender systems.

I remember being immersed in nature and movement as a child. People told me that I was a girl, and I used that as something to think about and with. It became a way of observing society and the world. Increasingly as I grew that assignment felt stifling, untrue, and inadequate. Movement was my way of feeling my own integrity. As I played and explored in grassy oak-filled hills and climbed gnarled and knotted pine trees in my neighborhood in the San Francisco East Bay, or ran joyously through sunsharp waves at Point Reyes National Seashore and kicked up the sand on the way home after dark as it sparkled with bioluminescent organisms, I could feel my own integrity. I knew in my heart that the girl stuff was not me, that it was something laid over me like a blanket, making it hard to breathe. That gender assignment became a way of understanding society as I grew into an adult and became more aware of the many intricate sticky webs of gendering that hold and define people. Being connected with nature saved me. It allowed me to feel my own self and source my strength in the face of a destructive alien onslaught. With each rock climb, hike, swim and paddle I could feel my own sources of knowing. The Two-Spirit and transgender authors featured here, Ivan Coyote (2012), Vivek Shraya (2016), Kai Cheng Thom (2019), Gwen Benaway (2016) and Joshua Whitehead (2017), also use gender as something to think with. Because they have been forced to question sources of identity and knowledge within themselves and without, Two-Spirit and transgender authors offer rich literary resources which can be used educationally to help students develop the kinds of complex, fluid, and relational thinking which we need today to address the ecological and cultural challenges that we face.

References

Benaway, G. (2016). Lake Michigan. In G. Benaway (Ed.), *Passage*, 3–4. Neyaashiinigmiing Reserve No. 27, ON: Kegedonce Press.

Brant, C. A. R. (2016). Teaching our teachers: Trans* and gender education in teacher preparation and professional development. In sj Miller (Ed.), *Teaching, affirming, and recognizing trans and gender creative youth* (pp. 47–61). London, UK: Palgrave Macmillan.

Carroll, R. (2018). *Transgender and the literary imagination: Changing gender in twentieth-century writing*. Edinburgh, UK: Edinburgh University Press.

Coyote, I. E. (2012). No Bikini. In Ivan E. Coyote (Ed.), *One in every crowd* (pp. 18–21). Vancouver, BC: Arsenal Pulp Press.

Ehrensaft, D. (2014). From gender identity disorder to gender creativity: The liberation of gender-nonconforming children and youth. In E. J. Meyer & A. P. Sansfacon (Eds.), *Supporting transgender & gender creative youth: Schools, families, and communities in Action* (pp. 13–25). New York: Peter Lang Publishing.

Fanon, F. (1967). Black skin, white masks. (C. Markman, Trans.). New York: Grove Press.

Fine, C. (2010). *Delusions of gender: How our minds, society, and Nnurosexism create difference*. New York: W. W. Norton & Company.

Ghisleni, L., Jordan, A. M., & Fioccoprile, E. (2016). Introduction to 'binary binds': Deconstructing sex and gender dichotomies in archaeological practice. *Journal of Archaeological Method and Theory, 23*, 765–787. <https://doi.org/10.1007/s10816-016-9296-9>.

Government of British Columbia (n.d.). *Sexual orientation and gender identity (SOGI)*. Retrieved from: <https://www2.gov.bc.ca/gov/content/erase/sogi>.

Grace, A. P. (2013). Camp fYrefly: Linking research to advocacy in community work with sexual and gender minority. In W. D. Pearce & J. Hillabold (Eds.), *Out spoken: perspectives on Queer identities* (pp. 127–142). Regina, SK: University of Regina Press.

Justice, D. H., Driskill, Q. L., Miranda, D., & Tatonetti, L. (Eds.). (2011). *Sovereign erotics: A collection of Two-Spirit literature*. Tucson, AZ: University of Arizona Press.

King, T. (1993). *Borders*. Retrieved from: <https://pennersf.files.wordpress.com/2010/10/borders.pdf>.

Lougheed, B. (2016). *The two-spirited collection*. University of Winnipeg Archives. Retrieved from: <https://www.digitaltransgenderarchive.net/downloads/w3763689t>.

Martin, R. S., & Walsh, J. (2021). *Man! I feel like a woman: Alexander the Great's exploration of gender*. Moving Trans History Forward Conference. University of Victoria. Retrieved from: <https://www.uvic.ca/mthf2021/program/index.php>.

McLeod, A. (2013). *Two-spirited collection*, box 1, folder 3–7, University of Winnipeg Archives.

Meyer, E. J. (2014). Supporting gender diversity in schools: Developmental and legal perspectives. In E. J. Meyer & A. P. Sansfacon (Eds.), *Supporting transgender & gender creative youth: Schools, families, and communities in Action* (pp. 69–84). New York: Peter Lang Publishing.

Mignolo, W. D. (2011). Geopolitics of sensing and knowing: On (de)coloniality, border thinking and epistemic disobedience. *Postcolonial Studies, 14*(3), 273–283.

Miller, sj. (2016). Why a queer literacy framework matters: Models for sustaining (a) gender self-determination and justice in today's schooling practices. In sj Miller (Ed.), *Teaching, affirming, and recognizing trans and gender creative youth: A queer literacy Framework*, London, UK: Palgrave Macmillan. <https://doi.org/10.1057/978-1-137-56766-6_2>.

Miller, T. (2019). Forget what you know about productivity and progress. *Yes Magazine*. Fall, 56–57.

Morgan, C., & Power, M. (2021). *Trans people in the Roman world*. Moving trans history forward conference. University of Victoria. Retrieved from: <https://www.uvic.ca/mthf2021/program/index.php>.

Rawson, J. K. (2015). Debunking the origin behind the word 'transgender.' *The News Minute*. Retrieved from: <https://www.thenewsminute.com/article/debunking-origin-behind-word- transgender>.

Roscoe, W. (1998). *Changing ones: Third and fourth genders in Native North America*. New York: St. Martin's Press.

Roughgarden, J. (2009). *Evolution's rainbow: Diversity, gender, and sexuality in nature and people*. University of California Press. ProQuest Ebook Central, <http://ebookcentral.proquest.com/lib/ucb/detail>.

Saskatchewan Ministry of Education. (2015). *Deepening the discussion: Gender and sexual diversity*. Retrieved from: <https://www.saskatchewan.ca/government/education-and-child-care-facility-administration/services-for-school-administrators/student-wellness-and-wellbeing/gender-and-sexual-diversity>.

Seymour, N. (2013). *Strange natures: Futurity, empathy, and the queer ecological imagination*. Champaign, IL: University of Illinois Press.

Shelton, S. A., & Lester, A. O. S. (2016). Risks and resiliency: Trans* students in the rural south. In sj Miller (Ed.), *Teaching, affirming, and recognizing trans and gender creative youth: A queer literacy Framework* (pp. 143–162). London, UK: Palgrave Macmillan.

Shraya, V. (2016). *The boy & the bindi*. Vancouver, BC: Arsenal Pulp Press.

Stryker, S., & Whittle, S. (2019). *The transgender studies reader*. London, UK: Routledge.

Thom, K. C. (2019). *From the stars in the sky to the fish in the sea*. Vancouver, BC: Arsenal Pulp Press.

Travers, A. (2014). Transformative gender justice as a framework for normalizing gender variance among children and youth. In E. J. Meyer & A. P. Sansfacon (Eds.), *Supporting transgender & gender creative youth: Schools, families, and communities in Action* (pp. 54–68). New York: Peter Lang Publishing.

Whitehead, J. (2017). The garbage eater. *Full-Metal Indigiqueer*. Vancouver, BC: Talonbooks, pp. 33–36.

Wilson, A. (2011). Coming in Native American Two-Spirit people. In M. Genovese & D.D. Russel (Eds.) *Safe and caring schools for two-spirit people: A guide to teachers and students*. Edmonton, AB: The Society for Safe and Caring Schools & Communities.

Epilogue

The narrative below is a threading of each author's "Love Letter to Other". As a method of decolonization, this threading is intended to demonstrate how we, as a group of authors in this book, depart from separate and discrete ways of knowing/being/thinking/doing/feeling, to take up a relational ontology. Expressing ourselves in this way, is to acknowledge that we do not exist as detached and independent from one another and planetary processes, but that decolonization is a complex layering of many different narratives entangled within assemblages of relations. While this narrative has undergone some edits for clarity, each sentence is derived from the author's original "Love Letter to Other".

A Love Letter to Other

Our Elders say everything has a spirit and is sacred: the white padded paws of the grey wolf gaiting through the enclosure, the sweet juice of a tomato as it explodes in the mouth, a first glance into eyes deep blue like ocean depths, the rosewoods of Northern Ghana rising above scattered savannah trees and decorating the northern vegetation with majestic umbrella crowns. Serving as a habitat for large bird species to nest and breed, the bound body slowly learns to give and take. We stand on the edge of what is holding us all together.

As Elders say, rosewood trees are sacred: protectors of the savannah and worshipped by the Kasenas. Every Northern ethnic group has a name for the rosewood tree because of their cultural significance. So is the sacred time waiting for the tomato. Starting the seeds indoors in a large window to shelter them from the bitter ice and snow, I wait since March. Or the sacred language of the fleshy, messy guts and bones of our vulnerability as living/dying critters, I thank the wolf for teaching me a different language. As a place to disconnect by reconnecting, I rethink who we are and where we came from. Attuning to the spilling laughter as children begin to play and chase each other:

> Hide and seek,
>> Wrestle and roll,
>>> Chew and lick.
>>>> Open, open to the world.

But why this reverence for rosewood trees? Or the admiration for tomato plants, as I watch in anticipation as tiny seeds sprout, watered every day until the last danger of frost passes in late May. Rosewood trees offer a door to the history of people and the story dates to the era of slavery. It is said that due to their thick crowns, the rosewood provided safe hiding places from slave raiders. As does the tomato plant provide a safe hiding space for Earthly critters. Rising from the earth to a glorious 6 feet, vines become laden with green fruit bathing in the late July sun. Yet as the fence of jagged edges between the wolf and me acts as a hyphen between wild and domestic, or without the romantic lens, perhaps just domestic and domestic, it is true that we are all tamed by capitalist and colonial dis/euphoria. As Indigenous peoples see the sacred, capitalism sees the resources. Wild secrets shrunk by globalised pathways of commercialism, are now just dots on a map. And sadly, the rosewood falls to fault. Coming under attack from the chainsaw, its beauty and strength mean it is an excellent choice for making furniture. In February I miss my tomatoes so much, I pluck a few from the grocery store, grown in Mexico, picked green by distant careful hands, and ripened in a refrigerator truck on the 3,000 km journey Northbound. It is never the same though. Left unsatisfied, I mumble to myself, I will just have to grow more of you this year.

Our Elders say it is our time now to protect the rosewood. Just like I carefully remove bottom branches of the tomato plant to prevent disease and pluck away other plants that try to crowd its space. I fuss over it. Does it need water? Is the ground too wet? Does it have everything it needs? Whenever I see you, I can see myself within you. Whenever I see you, I am inspired to ask how I am response-able to my relationships with you, others, and history. As the rosewood trees once protected us from slave raiders, we must now protect them from the chainsaw. Breaking the shell of custom,

> And letting the hot lava,
>> The beautiful breath,
>>> The essence out ...

As the right hand washes the left and the left hand in turn washes the right (an old adage in Ghana), in August I am overwhelmed by love. Eating my fill and sharing with friends, I stuff baskets, shirts, and cheeks full of deep reds, oranges and yellows. This is a place to feel free. Whenever I see you, it reminds me how I need to be becoming, unlearning, relearning, and reclaiming. I work over the stove for hours to preserve your goodness for winter, putting aside seeds of each kind to grow and share with your other lovers. I inhabit the ebbs and flows of your boundaries to explore my own human hybridity with the sky, the rocks, and the capitalized and colonial dirt in which we both stand. I now know that where there is power there is also response-ability. And I dwell in the fleeting moments on the Land. Surrounded by a community not my own, I have come to see that perhaps even more than the land itself, learning to live in reciprocity with the land is the greatest gift of all. You and I are "us". While I know the global story is not just from the Settler's pen, I am deeply sorry for the magnitude of this script. As settlers, knowledge of how to reciprocate the gifts of the land is rarely passed on, yet we fill with gratitude through commitments to (re)claim this literacy. And through this love affair, now reaffirmed, we begin a new cycle. Senses mixing:

>Taking in the sun,
>>The sand,
>>>The earth,
>>>>Falling downstairs,
>>>>>Dancing in the daylight.

About the Authors

Editors

Marcelo Gules Borges

Marcelo (pronouns: he/him/his) holds a B.A. in Biological Sciences, M.A. in Ecology (Environmental Sciences) from the Federal University of Rio Grande do Sul (UFRGS), Brazil, and was a visiting scholar in the Education Sciences Graduate Program at the Porto University in Portugal in 2007. He completed a Ph.D. in Education from the Pontifical Catholic University of Rio Grande do Sul (PUCRS) in 2014 having experienced eight months as a Visiting Scholar at the College of Education, University of Saskatchewan, Canada in 2012. In 2014, he was a postdoctoral fellow in the School of Humanities at the Pontifical Catholic University of Rio Grande do Sul (PUCRS) and has held a position in the School of Education, Federal University of Santa Catarina, Brazil since 2014, where he is a tenured professor. From 2018 to 2019, he was a postdoctoral fellow at the College of Education, University of Saskatchewan, Canada. He has been researching the anthropology of education, teacher education, and environmental and science education. From 2014 to 2017, he coordinated the project *Aprendizagem em redes: modos de habitar os lugares, as cidades e as escolas* [Learning in networks: Ways to dwell in the places, cities and schools] funded by Brazilian National Council for Scientific and Technological Development (CNPq/Brazil) and has recently published in the journal of *Environmental Education Research*.

Janet McVittie

Janet (pronouns: she/her/hers) is a researcher and teacher in the Department of Educational Foundations, University of Saskatchewan, Canada. Her research and teaching draw on her life-long love of being outdoors in places that have been less touched by Euro-American development. Her love of all things outdoors led her to first taking a B.Sc. in biology, and then a B.Ed.. She completed her M.Ed. in 1994, from the Department of Curriculum Studies, University of Saskatchewan, and her Ph.D. in 1999, from the Faculty of Education, at Simon Fraser University. Since then, she has taught and researched environmental, anti-racist, social justice, and sustainability education, publishing recently in the *Canadian Journal of Environmental Education*, the *Australian Journal of Environmental Education*, the *Journal of Environmental Education*, and the *International Journal of Early Childhood Environmental Education*, and a chapter in *Reconciliation in Practice: A Cross-Cultural Perspective*, published by Fernwood Publishing, in 2019. Janet has recently retired from her faculty position and is now working as a social and environmental volunteer activist with community and provincial organizations. Janet can be contacted at Janet.McVittie@usask.ca

Kathryn Riley

Kathryn (pronouns: she/her/hers) is a Postdoctoral Fellow in the School of Public Health at the University of Saskatchewan. Kathryn completed a B.A. in Education and a B.A. in Sport and Outdoor Recreation at Monash University in 2007. She then completed her Masters research through Deakin University in 2014. In 2019, Kathryn obtained a Ph.D. through Deakin University, conducting her research within the Saskatchewan-based education system, exploring new and different ways to (re)story human/nonhuman relationships for/with/in these Anthropocene times. Kathryn's research is focused posthumanist/new materialist scholarship that examines discursive (social) and material (matter) entanglements for social and ecological justice (in education). This has recently been taken up in scholarly publications for the *Journal of Australian Environmental Education*, the *Journal of Experiential Education*, and as a chapter for a book titled, *Outdoor Environmental Education in Higher Education: International Perspectives* (Eds. Thomas, G., Dyment, J., & Prince, H.). Kathryn can be contacted at kathryn.riley@usask.ca

Contributors

John B. Acharibasam

John (pronouns: he/him/his) holds a Ph.D. from the College of Education at the University of Saskatchewan. His Ph.D. research focused on environmental education, early childhood education and Indigenous knowledges. John's educational journey began from the University of Ghana, Legon where he obtained a Bachelor of Arts degree in Geography and Resource Development.

He proceeded to the University of Saskatchewan for a Master of Arts degree in Geography and Planning, focusing on strategic environmental impact assessment. He later transferred to the field of education through his academic experience at the graduate level where he was exposed to the serious disconnection from nature and the lack of environmental awareness in Ghana. This became particularly apparent in the midst of a wide range of Indigenous knowledges that have the cultural framework of love, respect, reciprocity, and responsibility towards nature.

Vince Anderson

Vince (pronouns he/him/his) recently completed a Ph.D. in the Program of Interdisciplinary Studies at the University of Saskatchewan, Canada. Vince's doctoral research concentrated on the integration of social justice and ecological justice pedagogies, particularly as it relates to community-based learning. Over the past ten years, Vince has instructed undergraduate courses at the University of Saskatchewan in Educational Foundations and Educational Psychology. He has also participated in a wide range of research projects related to providing support for teachers, with a specific focus on providing support to Indigenous teachers in Saskatchewan. Vince holds Masters and Bachelor's Degrees in Education from the University of Saskatchewan.

Ranjan Datta

Ranjan (pronouns: he/him/his) completed his Ph.D. in 2015 at the School of Environment and Sustainability at the University of Saskatchewan, Canada. He had already completed a Masters degree in Criminal Justice at Monmouth University, NJ, USA, as well as a second Master's in Sociology at Shahjalal University of Science and Technology, Bangladesh. Ranjan is of South Asian Indigenous descent from Bangladesh. He has a strong commitment to and passion for Indigenous environmental sustainability, Indigenous reconciliation, environmental management, Indigenous land rights, anti-racist theory and practice, decolonization, social and environmental justice, community gardens, and cross-cultural research methodology and methods. He has worked and advocated for the protection of Indigenous environment, land, and sustainability, particularly with South Asian and North American Indigenous communities. Within Canada, he is dedicated to building cross-cultural bridges among Indigenous, immigrant, and refugee communities.

Alice Johnston

Alice (pronouns: she/her/hers) is currently working towards obtaining her Ph.D., in Education at Queen's University. Her work examines how land education can be used to integrate Indigenous ways of knowing and being in the

K-12 curriculum in order to precipitate student engagement and decolonization. Before pursuing a Ph.D., Alice completed a Masters degree in Education from the University of Saskatchewan (2010), focusing on environmental and anti-oppressive education. Additionally, Alice taught for five years (2012–2017) as a middle years teacher in schools serving primarily First Nations and Métis learners. Alice worked for three years in La Loche, a Dene community in Northern Saskatchewan, followed by two years in an inner-city school in Saskatoon, Saskatchewan.

Roseann Kerr

Roseann Kerr (pronouns: she/her/hers) has recently received her Ph.D. in education from Queen's University, Canada. She has a Masters in Education from Queen's University (2009) where she focused on the perspectives of inner-city youth in out-of-school, experiential education programming. Rosie has a multidisciplinary background including studying and working in Fine Arts, Marine Biology, Curriculum Development, and Youth Leadership. Since 2013, she has been active in the food justice movement as a facilitator of community kitchen programs for Canadians living on low incomes. Her Ph.D. research focused on how *Campesino-a-Campesino* pedagogy supports Campesino/a empowerment and transition toward agroecology and community self-sufficiency in Southern Mexico.

Kai Orca

Kai (pronouns: they/them/theirs) is a Ph.D. student and Teaching Assistant in English at The University of Saskatchewan, researching the multicultural foundations of Canadian Two-Spirit and transgender literature. Their previous research and teaching has focussed on myth and mountaineering in American life (Ph.D. in Folklore, 2004), salmon and people on the Skeena River of British Columbia (project for the Royal BC Museum), and the anthropology of landscape and place. As President and fieldworker for the BC Folklore Society, Kai worked with loggers, dairy farmers, and storytellers. In addition, they spent time learning from Cameroonian farming women of Nso', Cameroon about connections between gender and land. From 2006 to 2018, Kai worked as a parent and educator of gender diverse children and began connecting gender identity with issues in environmental education. Completing a grades teaching certificate and M.A. in Waldorf Education in 2012, they have developed arts-based lesson plans and curriculum to help gender-diverse children become grounded and courageous individuals.

Index